Trauma and Lead

The Impact of Trauma on Leadership, Technology and Society

By Katrine Legg Hauger

ISDN: 978-82-93451-04-4

Part 2 The Quiet Evolution Book Series

In beginning:

"I am because we are." Ubuntu, Xhosa

"Authenticity is the alignment of head, mouth, heart, and feet – thinking, saying, feeling, and doing the same thing – consistently.

This builds trust, and followers love leaders they can trust." Lance Secretan

This book is dedicated to our dear children.

You are the future.

We are here to be brilliant.

Magnificent.

Let no limitations

and judgements lead you.

You are you.

And that is all there is.

To be the fully you.

Autonomous in

Togetherness.

Katrine Legg Hauger, 2012

What is leadership? What is trauma? What is a human being? What is identity – our own free will? What is impacting what – and how? Are you a leader – in your own life, parent, health worker, teamleader, social entrepreneur, storyteller, politician – or of an organization? Are you feeling overwhelmed? Being quiet, healing trauma, becoming connected and slowing down are needed now more than ever – and even more if you choose to be, grow, develop and play an autonomous and sustainable leadership role, whether in your own life or for a group of people. Deficiencies in our modern lifestyles of blindly day-walking through life - surviving without connection - goes against the grain of our nature. This is also the case for conscious and responsible leaders. Both we, our children and the people we are leading, are ravaged by survival strategies including stress, fatigue, infertility, addictions, violence, abuse, depression and confusion. Working on our own trauma affects our leadership capabilities and quality in a profound way, and reveals that most of us are too wounded to feel our underlying pain or to see the trauma in our systems. Our bodies' rejection of this not only causes our problems as individuals and leaders, but had the power to heal them – and to thrive as authentic leaders. Dare your vulnerability, dare your trauma, and thrive as a leader in your own life. This is the start of daring what is as it is, owning your story, sharing your unique gifts and staying grounded in the reality of the causes, challenges and understandings of your experienced human values in your body - for being daring and capable of seeing the greater picture in the organic body of any human relationship and group.

What if we didn't have to survive and could truly thrive our leadership skills? It takes one healthy human being to see, understand and support the healthy growth and development of another human being. The fact that we are often blocked from connecting with our bodies underlies most psychological and physiological problems and unhealthy hidden dynamics – in our lives, parenting, health systems, the utilization of technology and in our leadership roles, when innovating the future of our new generations. A unified approach to traditional psychology, modern trauma theory, systemic development and deeper leadership skills, integrates fragmented parts of our identity, psyche and the different hidden elements in our family or in our organization. These parts may include: early trauma, inner and outer victim-perpetrator dynamics, addictions, low self-esteem, stress, shame and self-judgment, while emphasizing our healthy resources, human leadership skills and resiliency capacities. This books step-by-step transformational insights and writing methods help you enter your quiet heart space for increased resilient leadership capacities, to fully explore the causes, consequences and perspectives in any system – for more fully understanding trauma and its impact on leadership, technology and society.

When we fully understand the impact of a healthy identity for healthy human development and growth - how womb ecology becomes world ecology - and see and experience the importance of our own free will and that the whole is greater than the sum of all of its parts - in our mothers womb, in our own bodies, and in any relationship and group - we will be capable of co-creating a humanitarian, sustainable and healthy future for years to come.

Katrine, Registered Psychotraumatherapist IoPT, Diplomized Systemic Coach Organization Constellator and Supervisor NKf, mother and lawyer, is the founder of Quiet Publishing Katrine Legg Hauger Int. and author of The Quiet Evolution book series.

Join us for The Quiet Evolution here: www.KatrineLeggHauger.com/.no

OVERVIEW

In this groundbreaking book, **Trauma and Leadership - The Impact of Trauma on Leadership, Technology and Society**, English Citizen but born and multi-culturally raised in Norway, mother, lawyer, author and Registered Psychotraumatherapist IoPT (Identity oriented Psychotraumatheory and –therapy) , Supervisor, Diplomized Systemic coach and Organization Constellator NKf Katrine Legg Hauger, describes how all people can become their best selves by healing multigenerational and early trauma, healing system trauma and emulating some of the great leaders of our time such as Gandhi and Martin Luther King Jr.; who remained true to their belief systems and resilient to oppression in order to serve as examples for others. It is distinguished by the belief that ordinary people are all capable of leadership; and that whatever form that leadership assumes (serving by example, writing, teaching, supporting, trauma work, systemic development and facilitation, storytelling or advocacy work) can make a significant contribution to affect change in the world.

This book is published after Hauger's book, **Trauma of Love - Quiet Your Mind For Healing Trauma To Shift From Surviving to Thriving**, (part 1 of a book series on human evolution and leadership), and provides readers with the practical tools to fully embody their leadership skills and become better leaders in their families, communities and in the world quite simply by healing trauma and becoming more in touch with their authentic selves. Hauger taps into proven techniques including: alignment with the latest science within deep trauma healing, the evolving consciousness of humanity and the crumbling of many of our modern paradigms (including technology, economics, social justice, law, mental health and politics); finding and embarking upon their true path; finding their quiet heart space, creativity flow, owning their leadership weaknesses in order to transform them into strengths, by joining forces with others and practicing systemic knowledge, development and innovation, compassion, resilience, deep listening skills, inner connection and interconnectivity in their daily lives.

She describes how the world has reached a phase where all people are being called upon to consider an integral humanitarian approach and given the tools to lead ourselves and others. She believes we are fortunate to be living during a time period where we are poised on the precipice of these changes; and have been given more tools to contribute in a more comprehensive fashion than we ever have before been given. This book will help readers use those tools to become authentic grounded, heartful and evolutionary leaders, peacemakers, social artists, storytellers and authentic social entrepreneurs.

There are few books on the market which combine the knowledge of

our shifting systemic and evolutionary realities and trauma healing with leadership; nor those that recognize that every individual has the power and responsibility to assume greater leadership roles and become vision bearers for heathy autonomous and responsible global change and sustainability. This book is accessible to everyone assuming leadership in any aspect of their life from business leaders to parents to office workers to authors. It is distinguished by the theme that everyone has the potential to step into greater roles and deeper understandings of their own realities and deeper human levels to affect change.

Hauger's book also touches upon some of the basic elements of conscious awareness, ownership and economics that contribute towards humanitarian goals. However, her book takes these themes further by outlining clear cut methods, exercises and practices readers can incorporate into their lives to step into greater leadership roles. Her memes are like: Heal Your Trauma. Own Your Story. Get Connected. Change The World. Her concept of One Globe Leadership™ is about aligning with the we space of co-creation, systemic interaction and guidance, moving from competition to collaboration for thriving on change and complexity, evolving social entrepreneurship and innovations, while emerging a new renaissance in leadership for a resilient economy and a sustainable humanity.

AUTHOR BIO

Katrine and her Quiet Publishing Katrine Legg Hauger International's mission is global new thought leadership, activism and peacemaking, for emerging a sustainable economy and future of our humanity. Her goal is to inspire you to embody the new story and tell the new news. Together we heal our early traumas and trauma of Identity for developing our own free will, catalyze, evolve and co-create inner and outer peace, connection, systemic development and wholehearted individual and global sustainability, health, freedom, abundance, social innovations, peacemaking and evolutionary leadership.

While being a just turned 41 years old grateful mother of two wonderful children, wife, leading and teaching psychotraumawork. while being a transformational Storymedicine™ authorand fulltime lawyer, she dreams of an awaken humanity igniting us all with the original flame of grounded authenticity, creativity and interconnectivity. Katrine is transcending and supporting the emergence of life-altering shifts within epigenetics, neuroscience and identity oriented early trauma healing for connecting with our true identity and body, altering our concepts of consciousness in our own lives and in the world. She is an inspiration to everyone to embrace the most magnificent graceful expression of themselves through connecting with their true I, reality and real story, while contributing to the evolution of a resilient, sustainable, peaceful and non-violent conscious global community.

Katrine brings profound inside-out business transformation and profitability to humanitarians, social entrepreneurs and leaders by bridging heartfulness, consciousness and business, serving visionaries who want to leap fearlessly into their deepest level of being. Her thought leadership on integral enlightenment, systemic work, law, trauma healing and mental health guides conscious leaders, individuals and professionals to find their healthy "I" in their own body, develop deeper self-worth, connectedness, autonomy, clarity, courage and direction to thrive in the new economy.

Katrine has helped clients make quantum progress through her unique combination of systemic thinking, deeper psychological theories like bonding and attachment theories and multigenerational psychotraumatology, the latest scientific neuroscience and modern trauma healing approaches, like the method of Constellations of the Sentence of the Intention (CoSI), developed by Prof. Dr. Franz Ruppert, based on your intentions, and her compassion, love, respect and non-judgmental professional approach. She is founder of The Traumalaboratory™ and The Quiet Evolution™ Movement, an expert at discovering and releasing blocks and traumas allowing clients to get back into the flow of thriving in their life, leadership, work and relationships. The clinically-validated powerful techniques Katrine uses are applied in a timely and sensitive manner, healing acute and life-long patterns of early trauma and emotional pain, abuse, violence, struggling, ancestral family-burden, stress, loss of loved-ones or systemic imbalances. Katrine helps you understand, integrate and resolve your most difficult problems in an entirely new way, expanding understanding, healing and restoring peace of mind within ourselves and in any organic system.

Katrine is a motivational visionary leader, incorporating individual trauma heling through the theories of bonding, and multigenerational psychotraumatology, collective field theory and embodiment with a cross-disciplinary philosophy for individuals, business, systems and economics. Katrine invites and inspires people to discover and embrace the most magnificent graceful expression of themselves. The more vulnerable and daring you are as an individual and a leader, the further you can reach in discovering and owning your real story – becoming an autonomous and responsible leader in all areas of life and business. She is guided by her own personal and professional story, connectedness and deep knowing that deep healing and greater change is possible. Over the years her quiet inner guidance has led her to curiosity, exploration and a desire to integrate and embrace her reality and autonomy and open up to all there is as it is. Katrine guides you to let down your internal barriers at many different levels. The barriers that have been recreating our story, mostly the stories told us by others. Katrines mission is supporting you to wake up to the reality of the seeds and deeper roots of your life. Your I.

Katrine empowers humanitarian teachers of all types. Katrine has a special interest in teaching women from all walks of life, to transform the traumas, violence and pain of the past, into a powerful present for a sustainable future. Our planet and humanity-at-large has never been in such need of skilled, trained and adept soulful women leaders. Women who can own and birth their own reality, stay healthy in their relationships, embrace their greatest leadership potential, and lead others into their true calling.

We see the world faces many crises. However, never before has women had the resources and freedoms to wake up, and consciously evolve ourselves and the world around us. Katrine believes the time is now, to empower conscious, powerful and brilliant women, and also men, to help midwife and give birth to a healthy balanced, grounded and sustainable livelihood. Katrine is passionate about contributing to the evolution of a humanitarian community. It all starts within – by opening up for our own healthy I, daring our vulnerability, self-healing, self-love, compassion and hope. Be, express and let yourself shine like the star you are. Just by truly being who you really are. That's how you transform yourself and the world.

Katrine is a published #1 bestselling author with her chapter The Alchemy of Connection – Through Life, Death, and Rebirth: At the Heart of Challenge Lies Grace – Embrace Your Sound Of Silence and Rebirth the Language of Your Soul, compiled by Christine Kloser in the book Pebbles in the Pond Wave 2 Spring 2013.

Katrine is educated within psychology and law, working full time as a customs and excise border management lawyer, finished her Cand. Jur. Law Degree in Oslo 2000, and her LL.M. in International, Commercial and European Law from University of Sheffield 2001. She is a Diplomized Systemic Coach Leadership and Organization Constellator DGfS/NKf (2012), a Certified Psychotraumatherapist IoPT DGfS/NKf (2013) graduating at the former Hellinger Institute (Now Institute for Traumawork) in Norway. She studied the 1. st and 2. nd 2-years Advanced International Training Program in Multigenerational Psychotraumatology/IoPT with Professor Dr. Franz Ruppert 2013-2014 and 2015-2016 for Certified Psychotraumatherapists, Psychologists, Psychiatrists, Doctors and professionals working in the field of trauma.

In earlier years Katrine graduated Generation One Intensive (2013) and Agents of Conscious Evolution (ACE 2012) with Barbara Marx Hubbard and The Shift Network, is a Feminine Power Graduate with Claire Zammit and Katherine Woodward Thomas, has studied with Marianne Williamson, Sage Lavine, Brene Brown, Craig Hamilton, Ryan Eliason, Jean Houston, as well as graduated MasterHeart within transformational writing with her beloved friend, mentor and co-author Christine Kloser.

TABLE OF CONTENTS

Chapter One: Introduction

We only have to look back to our history to recognize that the great leaders of the world, those who affected real change in the world, were all men and women who, on some level were more evolved in healing their trauma and daring their vulnerability, their real story—trusting their heartfulness through embracing their deeper emotional heart intelligence. Pick a leader, any leader. Mahatma Ghandi, Nelson Mandela, Martin Luther King Jr; all tapped into their own authentic selves and plumbed the depths of their deepest pain and real truths; to secure the strength necessary to resist oppression and empower themselves and others through non-violent methods.

Mogandas Ghandi for example, worked tirelessly for the end to discrimination of Indians under British colonization and the end of British rule. During a stint in a South African jail for opposing injustices against Indians living in that country, he read the book Civil Disobedience by Henry David Thoreau, which included theories and practices of non-violent protest which he incorporated into his work in India.

Martin Luther King Jr. launched major civil rights protests in the United States in the 1950s and 60s; during which he was beaten severely and arrested and jailed nearly 30 times. His campaign gained momentum in 1955 when Rosa Parks, an African American woman, walked onto a bus in segregated Birmingham Alabama, sat down in a whites only section and refused to give up her seat. King joined the movement which ended segregation on buses. His infamous "I have a dream" speech in 1963, which incorporated reality, compassion, resilience and hope to support his basic argument that all people, regardless of race, creed or color have the right to equal treatment and justice. His work and the speech which took place during a time when the country was already barely starting to shift towards a new way of thinking; shifted consciousness from a paradigm of lack and fear which affected individual and societal perspectives. In 1964 he was awarded the Nobel Peace Prize for his work here in Norway.

Nelson Mandela worked towards shifting societal perspectives and uprooting a well-established system of forced segregation in South Africa through non-violent means. He was imprisoned for 27 years during apartheid and treated miserable, beaten and tortured and even urinated on. Yet he maintained his steadfast internal belief system that what he was doing was right; and that change would occur and lived to not only see the end of apartheid but to become the nation's first black president and the winner of the Nobel Peace Prize.

When we look at these figures, we often look at the things they achieved and the obstacles they faced; and believe them to be superhuman, possessing qualities we do not have and could never hope to emulate. While it is true, that these figures walked different roads and had different destinies in store for them than we do; it is also important to recognize that we can learn to tap into comparable kinds of internal strengths in ourselves to resist oppression in a world where consciousness and perspective is shifting dramatically and all humans must work for change.

This book will teach you how to take the internal healing and searching work you have been doing as individuals to the next level to become authentic grounded leaders for in your own lives, your families, communities and, in turn, the world. It is true not every one of us is destined to achieve as dramatic results as the aforementioned leaders; to end apartheid or segregation or become peacemakers between two warring factions of our country. However, by tapping into our most authentic truths, we may naturally find that we have the power to help ourselves and then again others to create subtle shifts that make a real difference in our daily lives and in the lives of others.

The dictionary defines the word leader as a person who guides or directs a group and/or has the ability to lead. You may be naturally skeptical of the word leadership. However, if you look back towards your own life and think about all the times when you assumed "leadership" roles in regards to your co-workers, yourself, your friends, your children, we see that we act as leaders in all fashions of life – being alive means we are affecting others in one way or another – in healthy ways or in unhealthy ways. It's a matter of taking true responsibility for our own mental health and in our relations from conceiving a new human being to our last day.

This may happen quite naturally of its own accord; we may not even realize it is happening when it is happening; we may simply be serving as quiet examples or role models for others in our daily lives; our offices, on the street, in our living rooms. Others may recognize how we have shifted energetically through healthy growth, brought more autonomy, light and happiness into our lives; reacted admirably to challenging people; or simply were resilient, daring, responsible and open to face and see the reality with open eyes without any blindfold, when faced with a difficult or traumatic situation. Daring is the one who sees her reality with open eyes.

We may also choose to cultivate leadership in a deliberate manner. Odds are, if you have been doing some psychotraumawork, you may have already noticed that you have felt the impulse to take it one step further and to share your deeply felt and heard learned inner knowledge and experiences with others. We may seek to take on leadership roles that already exist in our workplace or families.

The internal work we have been doing might cause us to recognize we were previously on the wrong path for us; and to carve out new work, or creative projects or educations that naturally put us in positions where we act as leaders. Some positions associated with this may include counselors, advisors, health workers, entrepreneurs, authors, teachers and storytellers.

We may find ourselves yearning to take on roles we consider more important, but not be sure exactly how to do this. This is a wonderful position to be in in the world. The yearning, the impulse towards change or helping is in and of itself miraculous – it may be healthy help from our healthy grounded parts of our body and psyche and a mature autonomy where we have helped ourselves first, or unhealthy help of others out of fear of looking into our own trauma and real story, lack of a healthy I in our own body, or surviving into trying to help through unconscious manipulation tools and strategies, extremism of different kinds within esoteric directions, politics, religion or other belief systems, violent abuse, control or the illusionary thought that you are the savior of the world. If being a leader out of our healthy parts, it provides the canvas for creating something wonderful. There are many ways you can "test" out the form this creation may take on. We may volunteer to help out in homeless shelters, receive and support refugees like what we are experiencing in Europe in 2015 with the Syrian war, prisons, hospices, schools; apprentice with mentors we respect; audition for plays or bands or become docents at museums. We may come to tap into the deep listening skills we have acquired to recognize signs and people with whom we can align to accomplish our goals. We may learn how to pay attention to our body; to create subtle yet powerful shifts in our own perspectives and vibrational frequencies, to work through our old traumas and inner resistance for stating and committing to our healthy intentions and goals for our future. This book will inspire you and teach you a plethora of ways to learn how to walk down the new path you have embarked on in becoming the leader in your own life – one step after the other – for deepening your leadership skills in leading others in groups and organizations.

Another important aspect of doing this work, is that we come to recognize we are not alone; and tend to find ourselves gravitating towards like-minded individuals; people with similar perspectives on human development, growth and its effects on our deeper inner leadership skills ; inclined towards peacemaking and healing the world. In shedding unnecessary parts of our ego, in learning to trust others and being less isolated; we become part of something larger; a more collective, less lonely whole. We join forces. This naturally heals us, and many of the dissociations, neuroses, psychosis and bad habits that previously defined us, also in our leadership roles, fall away.

This collaboration may take many forms. We may tend towards collaborative projects through which we tap into collective resonance and intelligence that result in greater things than we might be able to accomplish alone. We may link up to form makeshift communities structured around common goals; such as farmers markets for healthy food or enclaves of economists, artists or writers. We may find ourselves aligning with movements that developed for the betterment of humanity; such as those to oppose racism, GMO farming, political or social inequities or abuse of animals or children. We may form social entrepreneurships, peace organizations, or "healing collaborations and partnerships" trading services such as systemic development or other forms of energy work with each other. We may work as peacemakers in our communities, advocating for strange or demonstrating methods of solving problems without violence.

Some of our impulses towards change is occurring because many of the aspects of our modern industrialized society is collapsing under its own weight. Economies no longer support life and the needs of the majority of people in our communities. People are finding themselves moved towards anger, or rejection of the old ways; to contribute towards a new kind of doing things that does not mirror all the pre-existing flaws in this difficult world, like collaborating for the safey of the refugees, supporting new science and promoting green economies, .

Some of our impulses towards change and leadership may be occurring because the world is changing rampantly and many of us are recognizing on a deeper level that it is our job to tap into this collective potential to circumvent catastrophe. Problems such as natural disasters, hunger, genocide and war are occurring at an unprecedented rate and are impossible to ignore. The Internet, radio, television, travel and radio are providing us with eyes and insight into all the places and ways in which people are suffering. As a result, we tend to recognize on a deeper level how much common ground we share with the people who are suffering and often triggering an impulse towards compassion and helping. The fact that we are reaching out to each other to do these things naturally; is also largely a result of the way we are evolving as human beings. It is not only that we feel compelled to help out of some false sense of obligation; but that we are meant to. We are hard wired to help, and helping actually makes us feel good and heals our own wounds and inconsistencies in character. We all know that we are dependent on bonding and healthy contact for life to sustain and grow. Separation is dividing while collaboration is building a sustainable future for us all.

In this book we will also touch on the theory of Heart or Emotional Intelligence in order to understand how much impact our own emotions and the emotions of others truly have in shaping our realities, perspectives and movements; how we may be impaired by emotions we do not know are affecting us and how to gain greater emotional maturity in order to guide ourselves and be supported by others. Accordingly, this book will teach us how to recognize, adapt to and heal core wounds; and to collaborate with others through a more emotionally mature, informed perspective – true leadership.

This is both a wonderful and terrible time to be alive. We are again realizing, as many of our ancestors did throughout history, that we need each other. In doing so; culture bearers, vision holders and social leaders are emerging within our communities. We are learning to recognize them;

and to follow their leads for inspiration, sharing, co-creating and guidance. At the same time, we are learning to be our own guides and to respectfully guide others; to recognize and support their talents and gifts.

Another result that living during this time is that the world is reconfiguring itself. At the same time, we are learning to sense and trust in wisdom that is not necessarily spelled out. First and foremost we must heal our early trauma so we can trust ourselves out of our healthy I in our own body. Then, we must learn to trust others. Next we must learn to acknowledge that we are leaders in our own lives, in our family when parenting, and of our organizations or businesses, Finally we must trust that change is possible. We recognize that as we embark on our own paths; serving as examples or advisors to friends, family and colleagues; as we form new enclaves and partnerships and embark on new projects; we are aligning with other groups of people all over the world doing the same thing. New organizations, communities and systems are emerging that are at once both advanced and primitive; like the systems of songlines the Aboriginal Australians mapped in order to communicate with each other, gather and hunt food and to survive. We are tapping into a new way of being that is advanced, yet largely undefinable; honing in on trauma healing, epigenetics, telepathic neuroscience, psychic abilities, intuitive talents and understanding; as we have suspected many animals can do. We are naturally evolving; building a new history and a new reality for ourselves and all future beings on this planet.

Chapter One:
Ways in which we are primed for new leadership in the world. Changing Tides and Self Awareness

There is no disputing the fact that we are poised on the precipice of great change in the world. On the one hand we are facing more terrorism, environmental;, political, religious and technological threats. In several decades large portions of populations could be contending with lack of access to resources such as clean drinking water, electricity and gasoline--- things many of us have come to take for granted. Global warming has spurred an unprecedented number of natural disasters monthly around the world; and threats to wildlife and plants that are integral to our survival as a species.

Technological advances also have allowed us to create biological, chemical and nuclear weapons that could wipe entire cities and towns off the face of the planet.

On the other hand, as societies, individuals and a global collective--- we are much more aware of our interconnection with each other and our need to consider other cultures and countries in the world in our decisions. The increasing popularity and acceptance, even advocating of psychology, neuroscience, epigenetics and psychotrauma healing in media, the classroom and within our own communities have allowed us to truly recognize some of the destructive chains we have been born into in our societies and families; to own our own problems, habits and reactions to them; and the ability to finally break free from some of these patterns so that our children are more liberated from them. The ramped up and widespread use of relatively new technologies such as the Internet reinforce our interconnection and ability to band together and help each other through difficult time.

We are in a prime position to create lasting change in the world by developing our leadership skills at a time when we need it most. Not only is the cultivation of individual and group leadership crucial to our survival; it has become possible for more of us to become leaders in our own right. We only need to look towards the qualities of great leaders throughout the world, to recognize the qualities we all share as human beings.

EI. Emotional Intelligence in Great Leaders in our Society

For years, intelligence was kind of compartmentalized. People often associated intelligence with IQ book smarts, academic learning; the processing of facts and formulas and associated mental gymnastics that lead to advanced degrees and high paying jobs.

In recent years, more credence has been given to something called Emotional Intelligence or IQ which is simply put peoples abilities to glean wisdom through processing emotions, or at the very least taking into account the emotions of self and others.

Emotional intelligence, also called intuition, has three major components.

The first is emotional awareness or the ability to own your own stuff; to identify and manage your emotional responses to the world. We acknowledge that a lot of the emotions that appear to be responses to what is happening in the "now" is actually "baggage" like early trauma, habitual emotional patterning created while we were still developing when we were very young children. For example in this society people often operate in a constant consistent state of stress or anxiety. Recognizing where this anxiety may be based in our childhood trauma or training, or others childhood or trauma is a prerequisite for emotional awareness, or even better, trauma awareness.

The second is the ability to gain a foothold in those emotions and where they are coming; and use (or not use them) when performing tasks such as critical thinking and problem solving.

The third is the ability to really regulate these emotions---as a response to self and others; and to use them to empathize with others, to be compassionate towards them; and to help them. This may be achieved through seemingly ordinary life events; such as calming down someone who is upset or helping someone to see a silvery lining in a dark circumstance. You may also be moved to empathize or be compassionate towards others in a more collective way that inspires long term activities geared towards change such as volunteering, advocacy work or joining a movement.

Trauma and Leadership

We will come to recognize that emotional intelligence is the base root of all great leaders in the world – many of them have integrated and healed their trauma and transformed their trauma into great resources of deeper respect, greater understanding of human values, patience, trust and hope.

In this book we will explore Emotional Intelligence in order to understand how much impact our own emotions and the emotions of others truly have in shaping realities, perspectives and movements; how we may be impaired by emotions we do not know are affecting us and how to gain greater emotional maturity in order to guide and be guided by others. Accordingly, this book will teach us how to recognize, adapt to and heal core wounds; and to collaborate with others through a more emotionally mature, informed perspective.

(psychology today.com)

Emotional Intelligence in Great Leaders

We only have to look back to our history to recognize that the great leaders of the world, those who affected real change in the world, were all men and women who, on some level were more daring, honest, authentic or evolved—and had higher degrees of emotional intelligence.

Pick a leader, any leader. Mahatma Ghandi, Nelson Mandela, Martin Luther King Jr; all tapped into their own authentic selves and plumbed the depths of their personal truths; to secure the strength necessary to resist oppression and empower themselves and others through non-violent methods. Think of how they may have been affected emotionally by all they had to endure in their quest to help others.

Ghandi for example, worked tirelessly for the end to discrimination of Indians under British colonization and the end of British rule. During a stint in a South African jail for opposing injustices against Indians living in that country, he read the book Civil Disobedience by Henry David Thoreau, which included theories and practices of non- violent protest which he incorporated into his work in India.

Martin Luther King Jr. launched major civil rights protests in the United States in the 1950s and 60s; during which he was beaten severely and arrested and jailed nearly 30 times. His campaign gained momentum in 1955 when Rosa Parks, an African American woman, walked onto a bus in segregated Birmingham Alabama, sat down in a whites only section and refused to give up her seat. King joined the movement which ended segregation on buses. His infamous "I have a dream" speech in 1963, which incorporated spirituality to support his basic argument that all people, regardless of race, creed or color have the right to equal treatment and justice. His work and the speech which took place during a time when the country was already barely starting to shift towards a new way of thinking; shifted consciousness which affected individual and societal perspectives. In 1964 he was awarded the Nobel Peace Prize in Oslo, Norway, for his work.

Nelson Mandela worked towards shifting societal perspectives and uprooting a well-established system of forced segregation in South Africa through non-violent means. He was imprisoned for 27 years during apartheid and treated miserable, beaten and tortured and even urinated on. Yet he maintained his steadfast internal belief system that what he was doing was right; and that change would occur. He lived to not only see the end of apartheid but to become the nation's first black president and the winner of the Nobel Peace Prize.

When we look at these figures, we often look at the things they achieved and the obstacles they faced; and believe them to be superhuman, possessing qualities we do not have and could never hope to emulate. While it is true that these figures walked different roads; it is also important to recognize that we can learn to tap into comparable kinds of internal strengths in ourselves to resist oppression in a world where consciousness and perspective is shifting dramatically and all humans must work for change.

Chapter Two:
What will you learn from this book?

This book will teach you how to take the internal healing and searching work you have been doing as individuals to the next level to become leaders for your families, communities and in turn, the world. It is true not every one of us is destined to achieve as dramatic results as the aforementioned leaders; to end apartheid or segregation or become peacemakers between two warring factions of our country. However, by tapping into our most authentic truths, by understanding our psychology, finding the roots of our emotions and owning them, we may naturally find that we have the power to help others create subtle shifts that make a real difference in the lives as others.

The dictionary defines the word leader as a person who guides or directs a group and/or has the ability to lead. You may be naturally skeptical of the word leadership. However, if you look back towards your own life and think about all the times when you assumed "leadership" roles in regards to your co-workers—you will see you too are capable of being a great leader.

This may happen quite naturally of its own accord. We may not even realize it is happening when it is happening. We may simply be serving as quiet examples or role models for others in our daily lives; our offices, on the street, in our living rooms. Others may recognize how we have shifted personally; brought more light and happiness into our lives; reacted admirably to challenging people; or simply were resilient when faced with a difficult or traumatic situation.

We may also choose to cultivate leadership in a deliberate manner. Odds are, if you have been doing this work, may have already noticed that you have felt the impulse to take it one step further and to share it with others. We may seek to take on leadership roles that already exist in our workplace or families.

The internal trauma healing we have been doing might cause us to recognize we were previously too separated and on the wrong path; and to carve out new work, or creative projects that naturally put us in leadership roles due to a greater extent of inner autonomy and responsibility. Some positions associated with this may include counselors, advisors, teachers, authors, people working with children or in health professions and storytellers.

We may find ourselves yearning to take on roles we consider more important, but not be sure exactly how to do this. This is a wonderful position to be in in the world. The yearning, the impulse towards change or helping is in and of itself miraculous. It provides the canvas for creating something wonderful. There are many ways you can "test" out the form this creation may take on. We may volunteer to help out in homeless shelters, prisons, hospices, refugees, schools; apprentice with mentors we respect; audition for plays or bands or become docents at museums. We may come to tap into the deep listening skills we have acquired to recognize people with whom we can align to accomplish our goals. We may learn how to pay attention; to create subtle yet powerful; shifts in our own perspectives and vibrational frequencies, to reconfigure our lives to manifest our dreams with seemingly little effort. This book will teach you a plethora of ways to learn how to walk down your new path.

Another important aspect of doing this work, is that we come to recognize we are not alone; and tend to find ourselves gravitating towards like-minded individuals; people with similar perspectives on environment, economics, social work and leadership; inclined towards a more integrated and holistic human law system, alternative dispute models, and peacemaking in our own life and in the world. In integrating our lost trauma parts, shedding unnecessary parts of our unhealthy surviving strategies, inner victim-perpetrator strategies or ego, in learning to trust our healthy I, others and being less isolated; we become part of something larger; a more collective, less lonely whole. We join forces. We speak out. This naturally heals us, and many of the neuroses, psychosis, judging, dogmas, taboos, fear, shame and bad habits that previously defined us, fall away.

This collaboration may take many forms. We may tend towards collaborative projects through which we tap into collective intelligence that results in greater things than we might be able to accomplish alone. We may link up to form makeshift communities structured around common goals; such as farmers markets for healthy food or enclaves of artists or writers. We may find ourselves aligning with movements that developed for the betterment of humanity; such as those to oppose racism, GMO farming, political or social inequities or abuse of animals or children. We may form social entrepreneurships, or "healing partnerships" trading services such as systematic development or other forms of energy work with each other. We may work as peacemakers in our communities, demonstrating methods of solving problems without violence.

Some of our impulses towards change is occurring because many of the aspects of our modern industrialized society is collapsing under its own weight. Economies no longer support life and the needs of the majority of people in our communities. People are finding themselves moved towards anger, frustration, confusion or rejection of the old ways; to contribute towards a new kind of doing things that does not mirror all the pre-existing flaws in this difficult world.

The world is changing rampantly and many of us are recognizing that it is our job to tap into our inner healthy and unhealthy dynamics and this collective potential to circumvent illness, bad health or global catastrophe.

IS THIS A REPETITION OF WHAT IS WRITTEN BEFORE??

Problems such as natural disasters, hunger, genocide and war are occurring at an unprecedented rate and are impossible to ignore. The Internet, radio, television, travel and radio are providing us with eyes and insight into all the places and ways in which people are suffering. As a result, we tend to recognize how much common ground we share with the people who are suffering and often triggering an impulse towards compassion and helping. The fact that we are reaching out to each other to do these things naturally; is also largely a result of the way we are evolving as human beings. It is not only that we feel compelled to help out of some false sense of obligation; but that we are meant to. We are hard wired to help, and helping actually makes us feel good and heals our own wounds and inconsistencies in character.

This is both a wonderful and terrible time to be alive. We are again realizing, as many of our ancestors did throughout history, that we need each other. In doing so; culture bearers, vision holders and true leaders are emerging within our communities. We are learning to recognize them; and to follow their leads for guidance. At the same time, we are learning to be our own guides and to guide others; to recognize and support their talents and gifts.

Another result that living during this time; is that the world is reconfiguring itself. At the same time, we are learning to sense and trust in wisdom that is not necessarily spelled out. We recognize that as we embark on our own paths; serving as examples or advisors to friends, family and colleagues; as we form new enclaves and partnerships and embark on new projects; we are aligning with other groups of people all over the world doing the same thing. We are naturally evolving; building a new history and a new reality for ourselves and all future beings on this planet.

Chapter 3:
What kind of leader will you be?
Examining your personality traits and talents and how to best use those to create change and benefit others.

Personality types. Gauging how you have inspired others in your life.

Leaders are not only the Martin Luther Kings and Nelson Mandela's of the world. They are not only the people with the gift of gab, or the most charisma, or the ones who inspire thousands at rally's or while on the political circuit.

Each and every one of us is a leader in our own right. We serve in different ways. Some are advocates for those who cannot speak up for themselves. Others serve by quiet example, as teachers, nurses, mothers, lawyers and doctors. Some serve as inspirational leaders through their music, art, theatre, writing, blogging, poetry or athletics. Some serve as role models to others simply through their own inner responsibility to find inner resilience, grounded balance, or through hard struggles to "get out" of poverty, anxiety, capitulation, apathy or a barrio, of a bad marriage or abusive situation, of dependence on drugs or alcohol. People who go to court to stand up against abusers or rapists, who speak up against an injustice at work, who write a letter to an executive whose company is causing harm—are all leaders. We all have leadership capabilities that we utilize throughout our lives. In some ways they are integral to our nature, almost automatic. Yet, we can always support the flourishing of our inner resources .

The first step to becoming a skilled leader is to acknowledge what you have. It is not necessary to reinvent the wheel. We can work within the framework of our own personalities and unique stories, gifts and talents to expand our leadership roles.

Step One, Personality assessment.

The first step is accessing what that personality is and how you have inspired others in your life.

There are formal and informal ways we may determine our leadership path. One of the simplest things you can do is to draw on your own history, personality and accomplishments.

Take a piece of paper or your journal and try these simple exercises.

a) Make a list of all the people who have ever thanked you or expressed gratitude about something you did to help them in their life. This may have been expressed in a formal or informal way. Did someone for example thank you for providing them with guidance or love, for helping them through a tough time or for saving their life?

b) After each item write specific actions you took that may have caused the person to feel grateful.

c) Write down what kind of leadership qualities, professional, parenting or relationship role you were in when you were contributing to that person's life. Were you a teacher, counselor, therapist, colleague, daughter, neighbor, mother, sister, friend or member of a support group? Were you being a good listener, witness, supporter, cheerleader? Use any term that resonates for you.

d) Do not discount the value of what you have achieved.

e) Start to think about how you might draw on these qualities in the future for a different job, role or responsibility.

The Meyers Briggs Personality Type Indicator.

The Meyers Briggs Type Indicator was created by Katharine Cook Briggs and her daughter Isabel Briggs Meyers. The indicator was based on Psychologist Carl Jung's research to assess personality and psychological types and was originally used during World War II to help women entering the war workforce for the first time to figure out what kind of job would best suit them. The test was first published in 1962 and has grown in complexity over the years---and now contains four primary ways we experience the world: intuition, sensation, feeling and thinking; and which is our dominant trait. It helps us determine the way we process and regard our experiences; and how these preferences lay the framework for our interests, values and drive. It was further broken utilized to determine 16 personality "types."

These include whether a person is more introverted (internal, loner) or extroverted (external, joiner); sensing (using physical senses) or intuitive (using instincts); thinking (using logic and objectivity) or feeling (drawing on values and subjectivity) and judging (purposeful, structured) or perceiving (laid back, open to change). Using those structural guidelines we are able to assess which of 16 different personality types we are. People use the test to gain insight into themselves, assess what careers they may be best suited for, and to assess what kinds of roles they may have in the world. For example. One may be determined to be a visionary: bound to break the rules and pioneer new routes. Another may be more of a supporter, helping to guide others to achieve great things or to heal from trauma.

The Meyers Briggs test is available free online. You may try it for a more structured analysis of your personality type. It's fun.

a) Owning your story to know your true essence and real history for telling the new story.

After you have found your story, own it. If you are like me and our tribe of storytellers, leaders, teachers and vision holders, you want to make a difference.

Maybe you feel this is a too big question? Maybe you think you are too small or isolated to have great impact.

As a leader, visionary, entrepreneur or small business owner, you know that to be able to be a social entrepreneur, you need to integrate your business with the holistic heart-opening truths – your reality. In this way you will truly not only be, butmake a difference in the world by your sole being. Your heart will deepen in knowing its own truth and your body will heal to build more healthy strength and energy.

My hope for you is that I can be a vision holder for you through The Quiet Evolution book series and movement, so you can more fully become inspired to trust yourself, your visions, intentions and expressions. Not only by writing, but by connecting authentically in every way with your inspired innate powers, hope and light. Maybe, little by little, if you keep this in mind, you can approach your true visions, messages, purpose, story and the alchemy of your heart more easily?

Are you a storyteller? Your story is uniquely yours – but it is also part of a collective story due to the fact that our ancestors and we all have lived through and experienced the same centuries and times with its' specific rules, belief systems, stages in developments and structures of the society. You can trust your story is needed to be told – and that many with you have the same experiences. This is particularly the case with the hard and difficult stories carrying the hidden secrets or deeper trauma in our generations or lives – like sexual abuse, violence, death penalties, inhuman structural and societal systems – our quiet realities and truths – the stories hidden due to taboos. Do you dare your story and your reality? Make your dream become a reality. We can all tell the story of our unique and collective evolution by the way it exists in and effects every human being. If we listen carefully, we can all approach, integrate, express and share the story of our deeper reality, through the language of our bodies, through a pencil.

We can all change the world with our stories. One story – one vision – one message – one heart at a time.

a) *Recognizing your unique gifts and drawing on your strengths and talents.*

Try another exercise in your journal. Write down all the talents and gifts that you know you have, or others have told you about. These could include obvious gifts such as an artistic talent, or less obvious gifts such as your ability to be a good listener, to listen with your heart, to help others to share their stories or to provide comfort during tough times. Next to each of these gifts write one or two ways in which they have come out in your life. For example. I am a good listener. Marianne told me her story this afternoon. You may also include less obvious gifts such as the ability to discern birdsong, a tune on the piano or tell when a storm is coming by the shifts in the air. Write down anything you feel brings you closer to your inner qualities, even gifts that are unique to you that you may have first noticed during childhood.

b) *Honing your authenticity and skills.*

Pick one or two events when your personality traits or leadership roles you assumed in your life shined through. Take a few deep breaths. Think back and try and remember, on a visceral level, from your most authentic sense of self what it felt like to be your best self during that experience. Where did you feel most affected in your body while you were having this experience? Did you feel it in your heart center, in your arms or shoulders? Remember that feeling. Sustain it for as long as you can. This may behow it feels to be your most settled, most authentic self.

Hold onto that memory. You can draw on it anytime when you want to build up your leadership skills.

c) *Purifying your intentions and mission statements*

Set an intention. It is important to be as clear as we can about what we want. Pinpoint what you want to do with what you have. The intention could be further developing the leadership skills you already possess and want to integrate more fully into your family life or workplace. You may seek to use leadership skills to step outside your comfort zone and assume a new role at a job, volunteer position or advocacy organization. Allow yourself to be vulnerable. Embrace your inner power and set the bar wherever you want. Just be clear. Set down your intention. Write it down on the paper. Speak it out loud. Decide for having constellations of the sentence of the intention to integrate the hidden trauma behind your problems with relationships and bonding, overdoing, stagnation or lack of energy or success in your leadership role – in your life.

d) *Sustaining leadership. Leading others starts by leading yourself.*

We know by now that it is important to take care of ourselves before we can take

care of others. The same is true of leadership. We cannot steer others on the right path, until we have our feet firmly planted in the ground there ourselves. We have to take responsibility for our own emotions, actions, spirit; we have to lead ourselves before we can lead others. Sometimes leadership occurs when we tune into our own authentic selves by seeing, acknowledging and embracing old traumas, inner splits, diverging inner dynamics or ancestral baggage through constellations of the intention work. Sometimes it occurs when we physically move to a place that is more comfortable for us, or clear out our space, or get rid of some extra responsibilities, inner clutter that is bogging us down. Sometimes we just have to attune to what our leadership abilities are, to set our intentions and take a leap of faith that everything is going to work out once we take action. Sometimes we need to learn to be more patient. When we have determined we are ready, close to anything is possible. As autonomous leaders we already know there are healthy limitations due to the fact that we are all human beings.

> e) *Determining what kind of leader you want to be. Quiet leader by example, public speaking, joining movements, artistic leader, peacemaker, change agent or leader in your community or family.*

There are all kinds of leaders out there whose examples we could follow. Many people grew up without role models or heroes. One of the greatest gifts of being an adult during transitional times is that we really have the ability to go back and truly see and understand our real story, the body language of our bodies, to recreate our realities both as individuals and as part of a collective and global community.

One good way to do this is to find people who inspire you. Sometimes this is simple to do---we know right off the bat who our heroes are. It may be someone famous we emulated during childhood. It may be someone non famous we knew during childhood---a relative, friend, someone in the neighborhood or village where we lived.

If you can't think of anyone right away, do some research. Think back over the years. Were there leaders who passed away, whose death affected you unexpectedly? Look through the news, old books. There are no limitations to who can be inspirational to you. It can be a world leader, an ice skater, a singer, a small business owner, that woman in the grocery store who always smiles at people, that man who defends animals. Odds are you will find one or more people who you admire.

In your journal, write down the names of these people. Leave plenty of room and write the things about them that inspired you most, what leadership qualities they possessed, what acts they performed in which these qualities shined through.

See if you can match any of the qualities of these people with qualities you determined you had in your own personality assessment. What kinds of things did they do to take these quality further? Make some concrete goals for yourself about what ways you might advance your leadership, based upon their examples.

Are there any ways you can ascertain to take a few steps closer to your

leadership role? Are there any leadership qualities that aren't necessarily endemic to your nature that you would like to develop and expand on in your life? Perhaps you are someone who serves by quiet example by nature but would like to hone your public speaking skills. Perhaps you are someone who is a great orator and amazing at organizing and stirring people to action; but need to work on your listening skills or your patience. You may not be good at organizing, and better at writing or designing newsletters. Although it is good to start from your base nature; it is also liberating to step out your comfort zone and acquire some new skills.

Make a list of the qualities you admire but may need a little more work on. Write down some concrete actions you can take to integrate these qualities into your life. Don't discount your abilities as too far-fetched. At this point anything is possible. We are far more capable than what we generally give ourselves credit for. Remember practice makes perfect. It doesn't take long before a new skill we are working on becomes a habit.

Make a list of the qualities you feel are getting in your way of assuming more leadership in your life. Are there any ways you might be able to eliminate these from your life, or at least move them further out of the way or by integrating them so that you can liberate yourself from hidden trauma or obstacles. If it is a habit, how can you evolve your body skills, intuition, inner knowledge or train your mind to break it? If it is a circumstance (for example not having enough free time) is there some way you can gently restructure things so that more is possible?

It is important to note that we might also find inspiration or leadership qualities that we admire in movements, groups, ideals and not necessarily in people. What movements, ideals or groups do you identify most with? What movements, ideals groups would you like to become involved with? What are the collective qualities of people associated with these movements? What truly resonates with your sole being? Your healthy I?

Chapter 4
Psychotraumatology, birth trauma/trauma resulting from conception, the time in the womb, childbirth, early childhood trauma and how to apply leadership principles for healing.

a) Conscious vs unconscious realities

Many of us spend a lot of our lives believing we are living fully consciously aware; yet being driven, triggered, motivated by traumas, habits, patterns and beliefs that function on a largely subconscious or even unconscious level. We live in societies where many people live their lives without full conscious awareness. Accordingly, collectively, we are also functioning largely on a less conscious level.

Factors that may keep us living without full conscious awareness as individuals include multigenerational trauma, early trauma---such as violence, abuse, racism, birth trauma---addictions, physical injuries, mental health issues or some forms of oppression. Factors that may keep us living without full conscious awareness collectively include bonding deficits, lack of human contact, violence (genocide, war, rape), political oppression (lack of freedom of speech, restrictive movement/curfews), class restrictions (such as a caste system) or unequal distribution of wealth (gap between the haves and have not's). We may also be retraumatized and scarred on subconscious levels due to shifting collective perception, for example when the media inundates us with images of terrorism, nature catastrophes, smuggling of narcotics and pirate goods, or disease (such as ISIS, and Ebola) and people become scared on a deep, primal level; believing they or their children are not safe.

As Prof. Dr. Franz Ruppert notes, when we are consciously aware, we function better both in our own lives and in society. We tend to have better communication with others, be more effective in problem solving and our planning.

In order to become more consciously aware; we have to first become aware of what is driving us subconsciously---both as individuals and a society. When there is a trauma, we might seek to identify the events and acts that caused this trauma and recover the memories associated with it.

b) Human beings as unique---Trauma as scars

Human beings

The human psyche is remarkable. It transforms reality as it is, to subjective reality for the bearer. It is our main means of perceiving, empathizing, thinking, remembering, dreaming and feeling. It is our multidimensional survival, but it is also selective. This means there is the potential for perception to be limited, to provide us with faulty information, to be self-destructive and to escape into illusions and survival strategies that are not healthy for us.

Just as whales have unique tail prints that they leave in their wake when they slap their tails down on the water, that scientists can use to identify them; so does each individual have a unique life print stemming from their unique inner psychic structures stemming from their life experiences, their personal history and their genetics due to their epigenetic ancestral background. Our bodies remember all this today, and use immence amounts of energy to suppress and reject our own reality today.

Our collective realities can also have unique characteristics due to the time, cultures and countries we are living in, due to our countries history, due to our socioeconomic or political realities; or the greater socioeconomic or political standing of our country.

As we have noted elsewhere in the book trauma can occur on many levels which can scar us.

c) Multigenerational trauma. Violence.

Abuse or other forms of violence can be passed down through generations—in obvious forms such as physical abuse, verbal abuse, sexual trauma, rape on incest; or in more subtle ways such as inner and outer victim-perpetrator dynamics, prejudice and hatred.

Often the roots of this trauma are passed down from our ancestors, or several generations of our families without anyone being consciously aware of them. They may have started on a political level; for example an ancestor who was persecuted during a war or was forced to be silent or a persecutor—committing violence on others---may take his anger or grief out on his family without recognizing what he is doing. A woman may grieve over the loss of a parent and become depressed (anger towards inwards) and not realize she is neglecting her children. Children who are affected by this bonding deficit, symbiotic trauma and neglect may wind up being affected in various manners such as contending with addictions, aggression, suicidality; or neglecting or being angry towards themselves, their parents, their own children and others.

Multigenerational trauma can also take place on a collective level as well. Traumas passed down over generations can effective entire societies; as well as specific groups and individuals on multiple levels. Germans living in Germany may still be feeling the effects of the Holocaust today; African Americans may still be feeling the effects of slavery; the effects of the genocide in Rwanda will probably be felt by Hutus and Tutsis for generations.

d) Multigenerational trauma. Birth trauma

Couples may be affected if they want but cannot have children, if they suffer from miscarriages, abortions, failed IVF attempts, modern birth technology, or the death of children at an early age

Babies may endure trauma during conception, in the womb before they are born or after they are born if their parent does not want them, is dealing with grief from the loss of another child, is dealing with trauma from their own childhood experiences (incest, neglect, abandonment, violence), is dealing with addiction or confusion. Parents may not recognize they are subconsciously passing these traumas on to their unborn or recently born children. Caesarean sections, attempts to abort and kill the babies in the womb, violence towards the mother, both physical and psychical or drugs utilized during pregnancy, a mother's alcohol or drug use can also cause trauma. An unborn child, fetus, or baby may feel deep trauma of loss and existential trauma associated with the loss of an unknown twin during conception, pregnancy or birth.

These children may suffer from deep psychological trauma like symbiotic trauma, system bonding trauma, existential trauma, trauma of love and employ mechanisms such as splitting and fragmentation of her inner psyche and bodily structures in order to survive.

e) Leadership roles to help people deal with trauma

Leaders are not only those on the front lines of political and social movements, policy makers or speechmakers.

As human beings living in the modern world, one of the most powerful things we can do to create change in the world is to help each other to see – with open eyes. Counselors, therapists, psychotraumatherapists Psychotraumatherapists, social workers, and other health professionals work to support traumatized people to align their subconscious realities into line with objective reality. They do this by uncovering the causes and roots of trauma, thus bringing subconscious or unconscious events to consciousness so that these can be embraced and healed at the deepest levels.

On a collective scale, this work may also be done in other supportive capacities.

World leaders may try to bring to light collective trauma that resulted from war or segregation; to psychotraumatherapists, systematic coaches or certified constellators/traumatherapists who bring patients and clients individual trauma to light; to artists, authors, leaders, parents, musicians and writers who try and bring to light an inequity in the world.

Simply by becoming consciously aware of the roots of issues that are damaging us as individuals and a society; can move mountains. We are living during a time when people's consciousness is shifting both individually and collectively. This is due largely to the fact that people are uncovering and accepting responsibility for their own traumas, that society is making it not only acceptable but desirable to do so – to become more autonomous.

Bene Brown PHD, is an author and research professor who advocates for individuals to embrace their vulnerability; and admit the things that have shamed them; in order to transform the way we live, love and lead. Brown asserts that it is only through embracing our trauma and the things that have harmed us; that we can really learn to trust each other and align with each other to heal ourselves, society and the world.

Leadership may occur in less obvious ways in cases where there is no real known obvious trauma—or a lesser known trauma. People who are more inclined towards introversion by nature, may have felt traumatized or persecuted throughout their lives due to their nature. They may have spent time on the sidelines, felt like it was unsafe to talk during a conversation, railed inside because nobody listened to, respected or asked for their opinion. They may have spent a lot of their lives standing back and witnessing what is going on in the world, and honing unique skills in the process to help others.

Susan Cain, a lawyer and author, pinpoints one of these unique groups in her book "Quiet. The Power of Introverts in a World Obsessed with Talking." In her book Cain points out how many societies obsessed with action and talking and who do not tend to give much attention or status to those who are quieter, more observant or better listeners by nature---can benefit from the skills of these individuals. Introverts are great at bearing witness to individual and collective trauma, to being able to help support others in their struggles to understand these traumas, to pave the way for change by helping others learn to take a step back, and be better listeners and watchers in the world.

Again, the most helpful ways we have to shift collective consciousness is to know ourselves; to learn to understand what makes us function on a deeper psychological level, to learn how we are traumatized, triggered and react to conflict in our lives. Once we learn to do this, we can learn to check in with ourselves whenever it is necessary. There is no time that this is more necessary as when large scale events occur that may cause widespread societal trauma in all levels of our society.

Chapter 5

Victim/Perpetrator relationships on large scales, inner and outer victim/perpetrator dynamics

In the previous chapter Prof. Dr. Franz Ruppert showed us how believing ourselves to be the victim as a method of surviving trauma, can scar us in various ways. We may align with the perpetrators by clinging to them on an emotional level, seeking to protect them, excusing their actions, by idealizing them and by not seeing them as they really are. We may even unconsciously seek to "fix" them.

Entangling and aligning with perpetrators may also cause us to become perpetrators towards others, or towards ourselves. It can cause us to become self-destructive; or turn our anger inwards which can cause physical (chronic diseases) and mental health problems, suicidal thoughts, eating disorders, cancer, depression or other ailments.

Being a perpetrator can be just as wounding as being a victim. When one commits an act that hurts others they can suffer from feelings of guilt, shame or ostracism; or having a negative conscience.

Likewise, in order to cope with the damage one has done, defense mechanisms may be employed. The perpetrator may not be able to acknowledge the harm they have done to another person; he/she may feel confused, like eg when traumatized by sexual trauma, righteous and justified for the acts that he/she committed (in lieu of feeling guilty). The perpetrator may project and blame the victim or feel himself/herself as being victimized. He/she may also create a scenerio in his/her brains in which the acts he/she committed are part of a larger ideology in which they see themselves as the hero. An extreme example may be a dictator who justifies his/her work committing genocide on a race of other people; by building a scenario in which that race is inferior, dangerous and/or threatening.

Perpetrators on large and small scales

Perpetrators and perpetration come in all forms. A man/woman may be an obvious perpetrator as in the case of a murderer or rapist. Then there are other levels of perpetration. A predatory mortgage lender may convince people with bad credit to invest their life savings in homes they cannot afford.

There are less severe forms of perpetrating actions which may not even be viewed that way. Office workers who form an alliance against another worker, investing time and energy to gossiping about him/her or trying to get him/her kicked out of the office may be perpetrators. They bring their victim – perpetrator dynamics into their parenting, leadership, workplace or their organization.

Subjectivity. Labeling victims and persecutors

The label of perpetrator is specific to the person who is labeling him/her. Someone may believe another people to be a perpetrator that others don't. A wife may believe a husband who cheats on her to be a perpetrator. A teenager may view a parent who won't allow him/her to go to a party to be a perpetrator. In the case of the office worker, the gossipers may feel themselves justified in their actions because that worker may have done something to them they may view as persecution---for example crossed a union line or lowered their salaries by coming into the office for less money.

Becoming what we learned

Assuming other roles, victim/perpetrator split, Victim becomes persecutor, persecutor becomes victims

The roles we assume as victims and persecutors can also be passed on. If a child is persecuted by an abusive parent; and subconsciously absorbs the ideology that parent lays down, he/she may grow up to be an abuser/persecutor or a victim to his/her child. He/she may not necessarily be aware that this is going on. He/she may be numbed towards it.

People may be neither true victims or persecutors. Their personalities may waver between each role and attitude. Anger at being victimized may be turned inwards and the former victim may become a persecutor towards him/herself. This may come out in self-destructive or self-sabotaging behaviors, aggression towards others, confusion, distraction, addictions, over doing, apathy, depression and/or suicidal behaviors or actions.

This persecution may come out in less direct ways. A man who is yelled at by his boss, may internalize his anger and come home and yell at his wife, child or dog. He/she may view aggression/depression as a normal part of bonding, human contact and living relationships. He she may be motivated by feelings of manipulation, suppression, victimizing every situation, revenge or false atonement, as he /she may not know of anything else.

Another common reaction to victim/perpetrator splitting is the inability to establish normal intimate relationships and friendships. People may invest a lot of energy trying to lay claim to false illusions of intimacy or love. This is made more difficult if the person caught in the victim/persecutor tailspin can't come to terms with his/her past/role and/or find the ability to love and accept him/herself. Many children are abused in the way that they are used as buffers and protection shields between their parents – in the bed or psychologically.

Getting Out. How to break the chain of victim-perpetrator dynamics in our own lives and our collective lives.

False attempts

People may acknowledge that something deep in their nature is damaged, that they have been caught in the victim persecutor cycle, be acting on triggers/habits that stem from old childhood trauma and blind symbiotic entanglements; and attempt to find a way to break this cycle.

We may first attempt to do this in earnest, but on superficial levels without getting to the root of the problem. This may even led to new abusive acts – although well meant – like when health professionals and conventional psychologists don't find the real causes and traumas behind their illnesses, decides inhuman medical surgery or operations when not really needed, provide medical treatment with immense unhealthy consequences, or don't really listen to their clients and patients stories. This may include acts of revenge to attempt to annihilate the victim or perpetrator or rebellion---struggling to fight the perpetrator or victim without real focus or intention. We may also attempt to forgive the perpetrator or reconcile within ourselves; without really uncovering, processing or integrating our own trauma related to the victim/perpetrator mindset. There are forms of escapism that people may practice to attempt to "fix" their problems related to victim/persecutor trauma without really leaving or disconnecting from the perpetrators. These include inner splitting, physical or mental illness.

Although art and spirituality may ultimately be important, even crucial means of getting to the root of where your trauma comes from, and healing; people are also apt to "practice" art and spirituality on levels that do not get to the heart of their trauma. They may trick themselves or cheat themselves due to survival strategies into believing they are healed, or have moved above what happened through techniques such as positive thinking which may gloss over, bypass or allow them to truly avoid dealing with and integrating their hidden trauma and causes behind their symptoms. In these cases people may "fake it" and convince themselves they have healed, or can save the world, when in actuality they are just creating and exercising another elaborate defense mechanism (for healthy and valid techniques of using innate resources for healing see chapter----).

In general, when there is real trauma, writing, leadership development and spirituality are most effective healing techniques when used in conjunction with more formal valid deep psychological practices such as constellation work based on bonding and attachment theories and multigenerational psychotraumatology.

Effective ways to overcome our victim trauma dynamics and attitudes. Individual

One of the toughest things for people to acknowledge is that they have been a victim – and also probably a perpetrator. This often requires one reopen old wounds, often reliving/re-experiencing trauma based upon what truly reveals itself of the clients inner psychical structures for the client and the representatives through neuron mirrors through the representatives in constellations of the intention, in order to acknowledge and accept that one has been wounded. Major complication contributing to people's denial of old trauma is the fact that victims and persecutors often do a number on perpetrators and victims; causing them to "blame" themselves for what happens, to mask what happened so nobody will know, or to view what happened to them as being routine or normal. Likewise, victims and perpetrators often employ a number of defense mechanisms (which become habitual) throughout their lives in order to survive the trauma. These often operate on deep subconscious levels, and as time passes it becomes more and more difficult to see and get rid of them.

Victims must learn to admit how they feel on deep emotional levels about what happened to them. They may cycle through emotions such as rage, shame, powerlessness, sadness confusion; for a while. After a while, as we open our eyes to the reality, these emotions harm us less and our surviving strategies become less dominant in our lives. We learn to feel them, and when we know how and why we developed them, we more easily can embrace them. After acknowledging these emotions, even to the point of courageously re-experiencing them and even confronting the perpetrator; they may make us feel sad in the future but they can never harm or break us again.

After admitting victimization, people must learn to feel compassion for themselves and what they have endured. This is different from self-pity, which may be necessary to experience for a time, but can ultimately become crippling and cause us to remain in the victim role. It is more like self-empathy. Imagine how you would feel if you witnessed somebody going through what you went through, how you would want to make them feel that you cared about and could understand what they went through; and that you want them to heal.

Ultimately, learning to overcome the victim attitude requires one to move past their victimization to recover their personal power; and to create healthy alternative ways to interact in the world. This includes recognizing one's own basic goodness, establishing good boundaries and establishing methods for taking care of ones damaged self.

This also requires we acknowledge, accept and seek to alter ways in which we may be emulating perpetrators attitudes, behavior and becoming perpetrators in our own lives.

Effective ways to overcome our perpetrator attitude dynamics--Individual

It is equally, if not more, difficult to acknowledge ways we have acted as perpetrators. It is helpful if we view our actions simply, in context, and acknowledge that there is no clear black and white, right or wrong view in life in general. We all act out different roles, and we all have the power to rid ourselves of undesirable roles and attitudes.

The first way to overcome our perpetrator action is to acknowledge things we did that may have harmed or victimized ourselves and others. If these are difficult acts to acknowledge for example abuse; we might also acknowledge that it is an act of bravery to do so; and that just by acknowledging it we are shifting things, as we also pave the way for the healing of the victim we harmed.

We must also "feel it" the guilt and responsibility for what we did, our anxiety about acceptance. We must seek to find compassion, and then empathy for the victims. This require that we have the ability to have contact with our body and not to escape from it into our head or even further out.

We might take concrete actions to make amends with victims; in some cases offering an apology or compensation (monetary or in some other form). Prof. Dr. Franz Ruppert mentions that persecutors should resist urges to make a lifelong atonement for what occurred.

Collective Victim and Persecutor mentalities and how to overcome them through autonomous leadership

THIS IS HOW FAR I HAVE COME 22.9

Victim/persecutors on a large collective scale.

We have seen throughout this book, how we can unwittingly pass on trauma, baggage from our ancestors, parents, selves and children; how trauma can be a chain, that if unacknowledged can continue on and on.

Similarly, this chain can be recognized on a collective scale. When incest and sexual abuse become the norm in a society; violence is also bound to be turned outwards. These may include obvious forms of violence such as murders, drugs and human trafficking, or political persecution of a group. It may contain less obvious forms of violence such as economic violence. Current economic systems which are driven purely by human greed/selfishness, where people do not exercise integrity in their economic dealings, where exploitation becomes an acceptable norm (getting ahead at the expense of the other, get them before they get you mentality); are obviously forged by a group of individuals who have been victims of persecution and/or are prone to violence. The effects of living in a society in which violence and persecution are inherent on subconscious and unconscious levels are also obvious in factors such as the extreme depletion of our natural resources and the poisoning of our natural environment, extreme gaps between rich and poor/have's and have nots, civil unrest in the form of protests, school shootings, mass executions and general terror. One would expect if we were living in more equitable societies where people trusted each other more, if the victim/persecutor split wasn't motivating us, we would see healthier, more balanced results. Franz….notes: That the victim/persecutor split results in the traumatization of bonding systems, and entire societies which are dominated by trauma.

We can also see the residual effect of large scale persecution in extreme examples of political violence. Populations who have been surrounded by violence in wars or genocides may go on to become exploiters of other races or populations; particularly if they were persecutors when they were young before their brains had a chance to fully develop. Children who grew up with parents in prison, or in tough neighborhoods where violence is the norm; may go on to wind up in prison themselves.

d) Alternative ways to heal Collective Traumas.

When our psyches are split, as individuals and collectives, terrible things can occur. Human beings may be demoralized, terrorized, oppressed due to their race, culture, or socioeconomic status.

Traditionally, systems settle old scores. These might be the politicians attempting to reconfigure an old systems (for example attempting to install a new leader after a dictator has been overthrown, or to try and institute

48

democracy in a formerly communist society). Lawyers mediate disputes whether they be criminal, business or family (as in the case of divorce).

Of course, this brand of quiet stilness does not always work. When the US went into Iraq purporting to attempt to overthrow Saddam Hussein and set up a new democratic system for the Iraqi people, they failed miserably. This happened because they failed to take into account the thousands of years of Iraqi history, their cultural realities, or even their wishes. Men, women and children are sentenced to prison terms that essentially keep them out of society; yet they often recycle back into communities where they commit similar crimes again. Prisons abound with men, women and children who committed crimes stemming directly or indirectly from childhood abuse or trauma. In these cases, mediation is not effective.

Alternative conflict resolution methods abound. Alternative Dispute Resolution is an alternative means of people or systems in dispute to come to a resolution without going to court. Some methods include negotiation (where there is no third party), mediation (where a third party the meditator facilitates the process but does not have power to decide what happens to either party) and collaborative law—through which lawyers intervene with preestablished standards outside of court.

There are programs which are developed which go beyond alternative conflict resolution methods. Restorative and community justice programs are currently being created through which men, women and youth who have committed crimes are not sent to prison but are given the opportunity to make amends to the families and communities they hurt through service projects. India has a system called "Lok Adalat" or people's court through which elders help resolve conflict within communities. Several tribal communities in Alaska have similar village court systems through which elders hear cases and dole out community service sentences in response to offences or crimes.

The latter is more effective because it goes to the root of the problem, the heart of the community and acknowledges on a deep level the humanity, standing and motivations for both victim and offender. It helps to heal trauma for both the community and the offender---by delving into its core. System wide processes tend to undermine the humanity of the situation. When people are warehoused away from the places where they committed crimes; the offender does not have ample opportunity to heal or make amends; and the victim and/or victim's family does not have the opportunity to forgive.

Any leadership we install to help create real change in the world, needs to delve deeper; to get to the root of individual and collective trauma, to dig out the decay and detritus and to clear the way for new growth to occur. The most effective way we can do this is to apply constellation work and transgenerational psychology on both individual and collective levels. We must seek to bring to light what ails us, in order to transform ourselves and the world. This can pave the way for more effective, systematic leadership to occur.

These kind of techniques are necessary in the case of large scale trauma or systematic violence, for example the case of genocide. When World War II occurred, many Christians who lived in villages where the

Nazis were in power, may have watched or contributed to the murder of their neighbors. The Jews whose families were exterminated may have found it difficult not to blame the perpetrators, may have found it difficult to forgive. Yet generations later---Germans and Jews who lived through the war together are finding ways to reconcile. Forgiving perpetrators can have massive healing effects on entire communities, societies and on the world. It is only when we acknowledge these traumas, acknowledge the effects they had on our ancestors and how our ancestors unwittingly passed those traumas down to our grandparents, parents and ultimately to us. These traumas may not be easily traced back to the source. It may take us some time to understand that the physical abuse our father committed on us, the abuse that we then went on to commit on our children through verbal abuse before we knew any better; is rooted in the trauma/pain of our grandfather who was forced to kill people of a different race during a war. It is only through this acknowledgement, through this unrooting, that we can find the means to break the cycles of violence and truly begin to heal.

We may also stop blanket large scale atrocities from subconsciously affecting our interactions with others on subconscious levels. For example On September 11, 2001 two planes were flown into the World Trade Center, and thousands of people were killed. The media inundated us with images of the destruction, and later with images of Muslims in Brooklyn cheering which caused many people to associate all Muslims or anyone they believed to be Muslim with the bombing. Others strove to look deeper. They did not associate the terrorists who flew the plane with their neighbors or men and woman who walked the street in their cities.

When terrorism, war, genocide, hunger or natural disasters occur, or we are threatened with their future occurrence--it would be in our best interest to learn techniques to help ourselves and others disseminate information, to identify and/or alleviate blanket political or cultural wounding or scarring as it might be occurring. We must learn to understand how we function in society on the deepest levels.

Getting Out. How to break the chain of victim persecutor in our collective lives.

Victims healing on collective levels

A remarkable thing happens when people begin to acknowledge the "terrible, shameful" ways they have been victimized. They come to recognize that there are resources out there to help them, guidance about how to heal; and perhaps just as if not more importantly that they are not alone. When one person begins to heal, they often find others that are healing from the same types of wounds. Groups of survivors may be found--for example in the various forms of Alcoholics anonymous or narcotics anonymous groups (including groups for family members of alcoholics or drug abusers); incest survival groups

When people begin to heal from these things, individually, and in collective groups; they become more and more acknowledged publically. Years ago, there was very little public acknowledgement of the incest or molestation of children that occurs at a rampant rate in society. There was no mention of it on television, or in the newspapers. Accordingly, the court systems did not support children who were molested. Sentences for abusers weren't severe, abuse was difficult to prove, and victims were often "revictimized" by court/legal and hospital systems when they came forward. Children were often viewed as liars or guilty; and the abusers often were free to travel through society largely invisible, often recommitting their crimes on more children. Parents of children who were abused by their spouses or other adults; often did not know where to turn or how to regard what happened to their children. They sometimes viewed it as the child's fault.

Times have changed. It is now common knowledge that children are molested, that adults commit incest. More victims are urged to come forward. Television programs across the board (talk shows, news programs, drama's, reality television) bring light to the fact that these things happen, that they are wrong, and that people who know victims should come forward and report the persecutors. Likewise, there is more therapy and groups for children that have been abused/molested. The courts have more support mechanisms in place for abused children; abusers receive stiffer sentences. There are registry's where abusers must go on public record about what they have done; they can no longer go through society invisible. Therapy and other measures are mandated to try and ensure they will not harm more children. It is no longer an invisible problem. We have made huge collective leaps in acknowledging that abuse happens, that it is wrong, and that we should make efforts to eradicate it from our realities.

Sometimes, leadership is realized because we are willing to witness the trauma inflicted on ourselves or other people; to give voice to that trauma, to join ranks with others who have been victimized or support those that have been victimized; to step up and bring to light a problem that needs to be quashed from our reality---a massive change that needs to occur. When enough people join ranks in this way: when we uncover the unacknowledged roots of individual and collective victimization; when we set an intention to acknowledge and heal from this victimization as individual and as a collective; we are shifting consciousness, bringing the subconscious to conscious light, paving the way for real lasting change to be created in the world.

Persecutors healing on societal levels

In this world, as human beings it is sometimes easy to blame persecutors. It is tough to see persecutors as former victims themselves, to acknowledge that some of their behavior is generally learned.

We are living in a transitional time in the world in the sense that people who were born in the past few generations; are finally seeking to break chains of violence---a lot of which is rooted in ancestral and collective trauma from a time that has passed. This is the age of admission. Psychological treatment in many forms (including constellation work) is becoming normalized. People are urged to seek professional help, to deal with their "problems" and issues. Seeking help is no longer viewed as a shameful thing.

Likewise, we are being urged to admit our collective trauma; to urge persecutors to admit their actions publically and to make amends for them. In this way our collective consciousness starts to shift.

There are programs through which persecutors are humanized in ways they never were. Cycles of violence which lead to the mass incarceration of young people of color from inner cities are being acknowledged. People are acknowledging that mass incarceration leads to cycles of recidivism, and does not work to deter crime. Programs are developed in which families of people who have been murdered are finding ways to forgive the murderer. They are spearheading alternative restorative justice programs in lieu of punitive measures.

Likewise, atrocities that were committed centuries ago are being publically acknowledged. People who committed war crimes, helping the Nazis, Albanians, Nazis, Contras to brutalize, torture and kill people are being brought to justice. Watchdogs like Amnesty International have been established to safeguard human rights that were blindly violated in the past. A good example of this is the situation that took place to safeguard the rights of protesters in Ferguson Missouri during protests following the shooting of an African American boy by a police officer.

Global response and increased awareness of the problems of human trafficking and sex trafficking are also good examples of people holding persecutors accountable on a societal level and shifts in collective consciousness through attention to the issue. Media attention, legal attention and protests/movements have been focused on holding persecutors accountable and attempting to break down the intricate system and chain that holds human beings as slaves across the world.

Another good example of collective response to perceived persecutors is the Occupy Wall Street movements in the United States which began when people joined forces on September 17, 2011 in the Wall Street area of Zoccoti Park; to protest economic and social inequities in the world. The movement drew it's inspiration by Spain's anti-austerity protests from the 15-M movement. The movement continues today with various groups springing up in different parts of the country and the world to protest the extreme inequities in wealth distribution (1 percent of the country holding the majority of the wealth) and the extreme influences of corporations who are not always held accountable for their actions and who hold what is perceived to be too much influence on the government and media.

A Word About Collective Response to Trauma

Just as human beings can make "false" or superficial attempts to overcome trauma and victim/perpetrator mentality; so can groups through movements and protests if they are not paying attention. An institution or individual may be scapegoated as a "persecutor" and a system set up to be overthrown, without proper attention being given to the root causes of the problem.

On the flip side, movements, protests, and massive shifts in collective consciousness which result from increased knowledge of a system or group that has been functioning as a persecutor and victimizing individuals or groups; which functions in an integral fashion by taking account of the root causes of the problems, its effect on individuals and groups is miraculous and has the potential to create real and lasting change.

Leadership potential in revamping individual and collective systems perpetuating the Victim/Perpetrator mentality

The mechanisms, structures, slogans and policies we adapt during this time can really affect change.

For example, when terrorism occurs, people join forces and adapt new mechanisms to resist it. There are leaders who pave the way for this to occur. Again, we can look towards Malala Yousafzai, the young Pakistani girl who wrote a blog when she was 12 years old detailing her life under Taliban occupation, and advocating for girls to go to school. Although Malala was shot by the Taliban on a bus, and nearly lost her life, she went on to become an outspoken activist for girls and womens rights to be educated and became the youngest person to win the Nobel Peace Prize.

Slogans can also pave the way for change. The phrase "Je suis Charlie" (French for I am Charlie) was a slogan adapted after twelve employees of the satirical weekly newspaper Charlie Hebdo, were massacred. The phrase became widely associated with freedom of speech and the press and as a device to resist terrorism, oppression and threats to freedom of self-expression.

Chapter 6:
Understanding evolution from an economic/societal viewpoint

We are fortunate to be alive during a transitional time period, when the old systems and paradigms we relied on and believed in within our lifetime are beginning to falter.

Just as modern man may have descended physically from less evolved species and ancestors (ie Y-chromosomal Adam); so too can we evolve in our roles in society.

The Collapse of old Systems

The changes in our lives in the past centuries have been profound. The economy moved from hands on agrarian and village communities where people made or traded most of the things that they needed to live, to the modern industrialized society where work was compartmentalized. Slowly, people became removed from the items they use, the food that they eat, the clothes that they wear. Basic elemental natural resources were cultivated from one source (largely machine based) and passed on to another source to be processed before it becomes an item we use. Table salt is extracted from mines, then chemically processed to remove some trace elements and spiked with iodine. Chemical dyes replaced native grown indigo and are added to vats in factories to dye blue jeans. A single

59

car can be made on three continents, using the resources endemic to each place. In the past few decades, we were even further removed from the material world with many industries being outsourced to laborers in other countries; and the advent of the service economy. This economy is a somewhat abstract one, with people providing services based on market demand and systems that are not necessarily grounded in anything tangible.

Technology

The service economy transitioned further into a technology based economy.

Modern computers and the World Wide Web have changed our lives dramatically. Now, we are not necessarily only victim to the machine, but are in a sense becoming a part of it. We have increased access to informational sources about anything and everything we ever wanted to know from how to diagnose your illness to how to ferment food. Politically, social media gives us access to information we may have never glimpsed before. ISIS's beheadings were broadcast to the general public via YouTube minutes before any political or media body could intercept it. This circumvented traditional systems of media checks and balances that have been in place for years. The possibilities about what kind of power this kind of direct access could yield us in the future are limitless. Such access could bring on positive results—organized non violent movements for peace or to show support for someone who is imprisoned. It could also bring on negative results---access to public executions, child pornography or calls to violence.

60

On the one hand, this new technology allows us to become part of a larger community of people from all over the world; to become more global in our sensibilities and projections. We have more control over what we know. On the other hand, many have argued that people become too dependent on the technology: that our creativity, memory, intimacy and critical thinking skills can be compromised.

Our reliance on computers has changed dramatically just in the past

fifteen years. In 2000 economists reported that clunky computers were bad business for the majority of firms. Now economists are projecting that future computers can produce things such as bionics, nanotechnology and alternative forms of fuel and electricity.

Politics and economics

The world economy soured throughout the world in the first decade or so of 2000. Greece went bankrupt in 2011. Crisis hit close to home throughout the US and Europe from 2008-2012 or so; brought on by predatory lending schemes, corporate greed, the burst housing bubble and displacing everyone from teachers to bankers from their jobs and creating a domino effect with millions of people living off unemployment funds for years and unable to purchase goods to put any money back into the economy.

This crisis is believed to have ended. However, some economists predict that there are a plethora of new crisis that may occur on a global scale in the not so far off future, due to the precarious and transmutable shape of our lives.

Risk factors could include the interconnectedness of financial markets combined with political and social turmoil; a stock market bubble that could burst, global competition for scarce resources such as those needed to produce energy or gas, world poverty and war and the general greed of large corporations and downgrade in quality.

The possibilities this brings you.

On the flip side, the interconnection of our economy and world; combined with our ability to access it through computers and travel with ease; has brought things to a new playing field. More people recognize the threats inherent in our current economic system both through lessons we learned in the recent past and those that could potentially be in store for us in the future. Environmental and social leaders and indigenous groups have a stronger voice and are better organized through massive events such as earth day, and bio-regional councils which usurp any one countries geographical boundaries.

People are coming together collectively to find alternative resources such as energy sources; and to both restore local economies and to share with others in need. On a small scale, this is recognized in such innovative industries as car shares, community gardens and farmers markets, small scale co-ops which loan money to support local businesses, and innovative means of generating income such as renting houses as vacation spots to individuals such as through AirBnb. There is a movement for more sustainable and ethical business practices on a large and small scale. People are beginning to integrate volunteer work, group retreats and environmental awareness into their businesses all over the globe.

Chapter 7:
Opening to humanity to guide science, lawmaking and leadership.

a) Daring your vulnerability for true courage.

Often, people are taught to associate vulnerability with weakness. We are taught to hide our vulnerability which may include our fears, love and the tender places in our hearts to protect ourselves from being hurt. We don armadillo type shells in order to prevent anything uncomfortable from reaching us, from preventing our vulnerability from becoming apparent to other people in the world.

What is vulnerability? It is rooted in emotions that are uncomfortable to us: such as unhappiness, terror, lack of control.

The impulse to flee from these emotions may actually have physiological roots. Our nervous systems are hard wired so that these emotions may set off red flags in our bodies; often characterized by a movement towards flight or fight; we resist the emotions fervently so that we can remain comfortable. We utilize all kinds of hay-wired circus tricks to avoid the circumstances that may cause us to experience these emotions, or to circumvent these emotions once they occur. At their worst, these may include a host of addictions to "numb" the emotions, violence or aggression as a reaction to it or isolation to avoid coming into contact with others that may cause us to experience it.

This belief system is false. It takes a lot more energy to resist and fear our vulnerability than to embrace it.

The greatest changes often take place when we have no choice but to accept our vulnerability. This is because our vulnerability is the portal to surrender to our higher powers, to spirit, to our Gods and Goddesses. When we have no choice but to accept our vulnerability, we can let go and accept the good we are given.

Think of times when you may have felt most vulnerable. Perhaps this includes a time when you were mildly vulnerable such as when you were nominated for an award or when you were up for a new promotion at a job; to times when you were more vulnerable---when you were pregnant and about to give birth, when you were speaking your mind in the face of opposition, when you arrived at the scene of an accident.

Vulnerability also helps us to connect with our fellow man, and other human beings on the purest level. The closeness that develops between people that are in the most harrowing circumstances where they are at their most vulnerable to death or sickness or attack---such as natural disasters or wars---is unparalleled. This exceptional shared vulnerability often becomes so deeply engrained in people during traumatic circumstances---that they seek each other out long after the fact and share their stories on a level they rarely do with others who have not experienced these things. The bonds that are created among survivors of common circumstances are deep and beautiful. Likewise, survivors of similar types of trauma who did not necessarily experience the trauma together at the same time---often join support groups with each other or seek each other out in friendships or relationships---in order to share their vulnerability with each other and heal. Examples of groups may include survivors of sexual assault, survivors of childhood abuse, veterans associations, alcoholics anonymous or narcotics anonymous. It takes a tremendous amount of courage to admit this kind of vulnerability and to share it with others. It also takes a tremendous amount of courage to admit it to yourself---to deal with emotions and emotional baggage associated with this vulnerability and the trauma that caused it that may be blocking you from other channels in your life.

Although those circumstances are obviously extreme, vulnerability can help us to connect to other humans on equally deep levels in everyday circumstances. Anytime we walk down the street and smile or make eye contact with strangers, when we give money or food to a homeless person, when we speak up for somebody else at work or at school, we are daring our vulnerability and tapping into our courage and forging connections in the most genuine manner.

Likewise we may forge these connections through our vulnerability without even realizing that we are doing this. We are most vulnerable when we are our truest selves, when we speak our own minds or live with integrity. In this way we may act as an inspiration or an example to others. This happens often in our ordinary lives, and is one of the ways that we fulfil the contracts we made before coming down to this world, without necessarily realizing we are doing so.

Embracing our vulnerability can also be a source of deep personal joy, relief, inspiration. Resisting what makes us vulnerable requires a tremendous amount of energy. Over time, this level of resistance, and the defense mechanisms---addictions, bad habits etc---that we utilize to keep this resistance up can become toxic and exhausting. It can block good things from getting through. Surrendering to the vulnerability, admitting that we are not in control all the time---frees us. With that freedom, positive emotions can get through.

In our perceived weakness, lies our greatest strength. Leadership requires that we embrace our vulnerability to connect with others on the deepest levels. Having the courage to embrace our vulnerability also helps us to tap into and become our most authentic selves. In wiping the slate clean of all our conditioning that teaches us to fear the unknown and to fear emotions that we perceive to be uncomfortable---we open ourselves up to embrace the spirit of each other and the world. In doing so, we can achieve remarkable things.

Embracing vulnerability for leadership may occur on a simple, practical level. It takes a lot of courage to take the next step to further your career---or to believe yourself worthy of a higher position. It likewise takes courage to abandon a career route that is not working for you but provides you with financial security or routine that has become safe and/or habitual for you, or that your family or community would discourage you from abandoning. Making bold career choices that are true to your nature but require you to become temporarily vulnerable and/or take a leap of faith that you will land on your feet---requires great stores of courage.

You may also feel vulnerable when you attempt to change your social network; especially if you have some degree of social anxiety or feel uncomfortable around strangers.

Some people may find it takes courage and vulnerability to break away from established paradigms in their families or communities. This is especially true when pre-established prejudices exist against one group of people---such as those defined by race, socioeconomic status or profession. It is also true when a person makes contact with another who has been ostracized by a group. Crossing lines that do not feel right, makes one vulnerable. It takes courage to be vulnerable in this manner.

A larger form of this takes place when people stand up for their political or social convictions, or against injustices in the forms of movements, protests and other forms of advocacy.

There is no place that people are more vulnerable than in their relationships with family, lovers and children. It takes great courage to be with people who know your flaws and weaknesses, as well as your strengths. It takes courage to be vulnerable enough to establish intimacy with others. When we do so, we pave the way for intimacy in all the extended relationships in our lives.

Children, especially very young children know how to be vulnerable. They do not disguise or prevent their emotions from rising to the surface. This is one of the reasons we are both drawn to children, and often find ourselves doubting them or seeking to cover up what they have to say.

Finally, it takes incredible stores of courage to be vulnerable enough to go against the grain of society, to embrace evolutionary innate truths, perceptions and gifts we are given and to follow them where they may lead us.

b) Forgiveness and responsibility.

Forgiveness is tough. Often we carry around baggage from our childhood and adulthood which includes all the wrongs we believe were committed against us. Sometimes we carry around this baggage on a conscious level. At others we carry it around on a subconscious level and do not recognize it is there.

Anger and resentment towards others who we believe have wronged us can become toxic. It eats away at the soul, and can prevent us from trusting ourselves or others; preventing intimacy in relationships; and on some level preventing us from moving forward in our lives.

The things we carry may be real or imaginary ways people have wronged us. These things may be indisputably wrong: such as if a man murdered our sister, a father raped his daughter or a wife cheated on her husband. The indisputable wrongs tend to carry the most weight, and take real work, patience and practice to get rid of.

The things we carry may also be perceived wrongs, or those that are less obvious. We may for example carry anger towards a parent who favored a sibling over us, or who wasn't around enough, or a lover we believe is not giving us enough attention. We may not even realize we are carrying these burdens around. Yet, when we are, they often surface due to triggers in other relationships or paths in our lives and wreak havoc.

Forgiveness is a kindness you do foremost for yourself. Letting go of anger and resentment towards another person---even if that anger or resentment is deserved---can help you to heal. Traumas, old wounds can close up if the forgiveness balm is applied for a while. Forgiving others either in an obvious manner, such as in person, or simply forgiving them within yourself can free you of things that prevent you from healing or growing as a person, from relating or loving others, and other roadblocks in your lives. In turn, it may also help the person who you are forgiving to heal and prevent them from committing similar offenses in the future.

Forgiveness can be hard, especially in cases where serious wrongs were committed against you. However, there have been the most extraordinary cases of forgiveness, where people were able to get rid of anger towards other people committed during extreme circumstances. For example, a woman whose son was murdered visits the prison to visit the murderer or requests leniency for the murderer because she respects his humanity. Survivors of genocide have forgiven their attackers. Women have forgiven their rapists.

Forgiveness can require that we find sympathy, or even compassion for the person who we believe to have wronged us. Sometimes it is not possible to find these emotions for another person who has wronged us. It is still possible to let go.

There are extreme cases where people may be instructed not to forgive. Sometimes psychologists advise people who suffered repeated sexual abuse as children not to forgive their abusers, and are able to instead find ways of letting go of the anger and forgiving themselves. This is a valid technique for such harmful circumstances.

However, for the most part, forgiveness is something to strive for in all our relationships. When we claim responsibility for ourselves by forgiving others, we strengthen our being and potential. When we forgive others, we bring our own small bit of healing to the world. Leadership requires we become more human evolved human beings in all our relationships. Forgiving those that are a primary part of your life and development, will pave the way for healthy relationships in new avenues in your life.

c) The road to compassion is paved with healthy love.

Compassion is the ability to find empathy for another person's

suffering and to feel an impulse towards helping them to ease that suffering. The most important element of finding compassion for others is therefore healthy love---both self-love and the love of the world as a whole. The more work we do for ourselves to understand and accept ourselves and the place in the world, to dare to be vulnerable, to let go of our anger and resentment towards others---the greater our capacity to healthy love. When we are surrounded with healthy love, when we are able to tap into it through connection to nature and others---we tend to find more compassion.

Compassion is a force to be reckoned with. It brings forth change in our own development and the development of others. The inclination to help, which is actually an integral part of our nature as human beings and part of our natural state---can shift things on an evolutionary scale by inspiring great change in society and aligning us as a group with spirit.

We might strive for greater compassion in all of our relationships; work and social engagements. In doing so, we forge a greater connection with each other and with all that is and all that will be. We become both individually and as a group, as Ghandi instructed, the change we wish to see in the world.

Getting connected.

There is no greater gift you can give yourself than the ability

to connect; to others, to ourselves, to nature on an authentic level. Our modern industrialized society and world tends to induce isolation, compartmentalization and disconnection in order to insure economic and personal survival. Lately, over the past several years or more---we have begun to sense how unhealthy that way of living can be for people. We have seen this in obvious, society wide ways---as in the collapse of the housing market in the United States; and in more personal ways---in increased rates of addictions, neurosis, depression, psychoses and other psychological and innate problems.

It is only by surrendering to nature, to spirit, to all that is and all that will be---that we begin to really connect with the world and gain the tools to really connect with each other. When we connect with each other on a genuine, deep level attuned to the magic and spirit and need for change in our society and the world---we begin to affect change. We evolve as a species due to this deep connection.

We may recognize the potential for this kind of change---when deeper connections become easier for us and less of a burden. We may find ourselves connecting on a deeper level with colleagues, neighbors, members of a club or group or social movement---and feeling gratitude for this connection in a way we have not felt before. In doing so, we may recognize that we have been bereft of this kind of connection in the past.

d) From surviving to manifesting goodness.

From early childhood, our nervous systems have been wired towards

survival. Our habits and conditioning are deeply rooted in protecting ourselves from anything we perceive to threaten that survival---generally through the flight fight or freeze mechanisms.

Society tends to exacerbate the problems that may arise through these protective devices or defense mechanisms. Our corporate and litigious culture tends to advocate for character traits that are not necessarily kind towards our fellow human beings. Phrases that have been utilized in these cultures include: Get them before they get you. Trust no one. Fake it until you make it. Several modern corporate executives have recently proven that in order to achieve success through the old survival of the fittest type of mentality; dishonest, greedy and manipulative methods have been used. Even those who maintain integrity and honor in their personal and family lives may be swayed to follow business models which are self-serving in order to thrive.

Lately, of course, as our financial institutions collapse and more people have lost their jobs or become dissatisfied with their work; these old models have changed. People are seeking to bring more of their everyday ethics and integrity into their business realm, to forge authentic relationships and connections with each other and even to create businesses that also serve others through new business models, the sharing of profits with others, cooperative agreements within corporations and the greater good.

As humanity evolves, we will seek out and manifest more of this genuine goodness in society, family, community and in our ordinary lives. Likewise, people are coming to understand that the old paradigms, even those well established during childhood through habits we took on due to our survival instincts (defined by and passed down to us through our parents, teachers and other members of society) are not set in stone. We are coming to understand that all humans carry within them an innate basic goodness and that can be trusted as our core nature. We do not have to play a lot of fancy psychological tricks on ourselves in order to survive, because our survival in fact was never threatened. This "deprogramming" is allowing us to envision greater things for ourselves as individuals, a culture, society, country and world as a whole that are not handicapped by fear based thinking. It is allowing us to align more with the universal spirit and our innate way of doing things as human beings living in a fragile and evolving world.

e) The importance of generosity.

Some of the worst habits we have taken living in the modern

industrialized world is to be tight-fisted with our money, dreams, listening and healthy love.

We have been taught that being selfish about these things may be integral to our survival. If we give too much away, we are taught, then we will not have anything left for ourselves. We have been taught that the world's resources are scarce and we might hoard them in order to protect ourselves.

This is another false belief. In fact it is only when we are generous with ourselves, our resources and healthy love, it is only when we are sharing with others and serving the greater good; that we free ourselves to give and receive abundance. More of the best things in life become available to us when we

One of the easiest ways in which we can give to each other is free---listening. When we are not afraid we will be attacked, spoken over or ignored---we can listen to each other's stories on the deepest, most impenetrable levels. When we listen with our hearts, when we do not interrupt what others are saying because we feel compelled to share a similar stories or interject superficial advice, we will eventually learn to hear not only what the other person is saying---but the way that their story resonates with spirit and informs us in our own life. Deep listening paves the way for true wisdom, for insight we can offer to the person we are listening to that is real and grounded, that contributes to their knowledge of self and the world. When we offer advice, then, it paves the way for the individual to make better choices, to create new paradigms to hone their contribution to the world. When we listen deeply, we are generous with our souls and bodies. It is this generosity that can contribute to curing psycho-spiritual and physical sicknesses, and create much needed change in the world.

Generosity requires many of the traits we have already discussed in this chapter: vulnerability to give of one self and trust that one's power will not be taken away, becoming connected with others on the deepest level, compassion, healthy love and moving past survival and other defense mechanisms to embrace and manifest goodness in self and others.

The pendulum is beginning to swing the other way. Modern industrialized society has, in the past, tended to discount and downplay real generosity in lieu of other notions of being successful. However, as the world shifts, people are beginning to band together more and to embrace and cultivate generosity. This is evident by all the small random acts of kindness people post on the Internet---from enormous tips given to waitresses to coats being given to the homeless. It is evident by the social movements that have sprung up to serve others rather than self; by the increase in businesses who give portions of their profits away to those in need; by the rise in people volunteering. Generosity is now celebrated, largely because people are sensing on a deep primordial level that generosity, rather than selfishness is the key to our survival.

The reasons for this are most obvious when we look at the challenges we may face securing water, energy and food in the future. We have begun to recognize that if we do not work together, there may in fact not be enough to go around in the future. Likewise, the number of natural disasters, wars and other tumultuous events going on in the world has caused people to instinctually band together and help each other out. I believe we are doing a lot of this because instinctually we realize we are poised on the precipice of major change in the world; and that this change requires we are generous with each other in order to survive. It turns the old paradigm of selfishness being necessary to survive inside out, and allows us to return to the generosity that is actually more endemic to our nature. When we are generous, when we help others, physiological reactions that are actually good for our physical, psychological and health occur. Tapping into generosity can guide our leadership efforts and ensure we are in tip top shape in everything that we do.

Generosity is a kind of failsafe measure we can employ when we don't know what else to do. Have you ever been poor enough that you can't afford something you want, then decide to give someone else your last dollar? Would you give someone who was freezing the jacket off your back or share your only meal? Being generous in this manner paves the way for miraculous and magical things to occur both within our own spirits and psyches (through the act of letting go of fear and attachment to scarcity) and on a practical level in the world). Try it and see what happens.

f) Transparence. A key to increased consciousness.

Again, we do not have to hide our true selves from our own

consciousness or from others. It is easy to say this, and less hard to do. When we look at ourselves with complete transparency: we are also forced to look at all the false habits, beliefs and defense mechanisms we are utilizing on a daily basis. We are forced to look at where our own junk comes from. For example, if we are having trouble with an intimate partner in our life, if we are feeling an unwieldy amount of anger towards them that do not seem proportionate to his/her actions; we may have to dig deep within ourselves and recognize our anger is rooted in a childhood relationship or trauma and to deal with that. If we are yelling at our child too much, we may come to recognize that we are just mirroring the actions of our parents in the past. If we are drawn again and again to poverty and scarcity rather than abundance, we may recognize that is a learned behavior stemming from a physiological reaction to early poverty that has been drilled into our system. Becoming transparent with ourselves is often extremely complicated and difficult. However, it is only after we remove the blinders to achieve this kind of clarity, that we can steer ourselves in the right direction.

Transparency can also work wonders in our interpersonal relationships. Striving to be as honest and clear as we can about our intentions and in our actions; can help others to see us for who we are. There are simple methods for attaining transparency. The next time you meet someone new, try a little experiment. Be as fearless as you can when you greet them, open your hearts to them in the way you did to other children when you were a young child. When you speak with them, be honest, hide nothing, pretend you never knew what it was like to be defensive or guarded. Simply be. Be yourself. Try and sustain this mindset for as long as you can. Practice it with all your new encounters with strangers or new encounters with those you have known for years. It is amazing how easy it can be to remain calm, present and productive when you are operating with complete transparency.

g) Cutting edge science and leadership.

You know the old adage, necessity is the mother of invention.

I would invite you to consider all the things the world needs in order to heal and

prepare for the future and new inventions which may solve some of our problems. For example, as mentioned earlier, many of the natural resources we need in order to survive or to thrive may be in short supply in the future. Many of our electrical grids do not have enough juice to power our cities just ten years from now. Many of our rivers, lakes, streams and oceans are contaminated; and clean potable water is not now available in all parts of the world. Our food supplies are becoming more and more limited due to natural disasters such as drought and flooding; and man made problem such as agribusinesses which pollute land and displace small family farmers. Our gasoline supplies are limited, and have become the basis for war and other acts of atrocity around the world. Diseases are wiping out entire villages and sectors of the population.

Accordingly, scientists and other inventors have begun to go to town to try and solve some of these problems. Electrical plants have been developed that are powered by trash, fish parts, solar or wind power. New devices to clean water have been developed. In addition to protests that occur against agribusiness and factory farms, and small scale community gardens and farmers markets that have sprung up; alternative forms of growing food in extreme circumstances have been developed such as hydroponics and vertical indoor gardens that are nourished by air. Inventors have created cars that run on electricity, French fry oil, and even those that fly through the sky. Likewise, alternative forms of transportation such as tube trains are in the works.

Scientists across the board from geneticists to physiologists are working for cures to diseases ranging from HIV/AIDS to Ebola.

All of these efforts can be viewed in the right light as humanitarian efforts. Scientists are developing products based on being informed about the deepest needs people have now and in the future and are driven by the (either self-imposed or societal) impulse towards helping.

Likewise, all kinds of leadership efforts are developed due to the need to help society develop new adaptations to survive and thrive in the future. Social movements and other forms of resistance and protest make it evident that we are seeking change in the world. People have mobilized in protests against agribusiness through Monsanto, and in environmental efforts geared towards fighting global warming, pollution, the displacement of entire communities. Likewise, businesses have sprung up that are more conscientious---which seek to limit their impact on the environment, to hire specific groups of people who may have lost their power through circumstance (such as homeless people or veterans), to give back to communities in which they are placed. These efforts pave the way for other businesses to follow suit, for new businesses to be developed, for new regulations that help maintain standards of quality and integrity to be put in place.

(Katrine. I think you probably have some ideas in mind for the law sections, so I am just putting a little bit about these in there—let me know what you want to add/subtract)

h) Law and mental health. A question of integration.

Many of us fear the law. This is largely because many of us live in

a litigious, lawsuit happy society. Many professions from hospice and palliative care to construction to fishing have extreme weights and measures in place in order to protect themselves from future lawsuits. You cannot visit a doctor, or buy a car or volunteer to help people, it seems, without being subjected to filling out a copious amount of forms that protect the people you are dealing with against future lawsuits by you.

On the one hand, too much litigation is dangerous. People are diverted from doing their jobs, and from trusting in the basic goodness of others due to the demands of the law. On the other hand, of course, laws are necessary to protect us, to keep society balanced and in check.

Just as inventions can be created to help people, as a response to societal needs, so can laws. Laws can be ethical and they can be designed to help. In creating news laws or accessing new ones it is important that we place human rights, dignity and needs at the forefront of our efforts. It is important that we empower people rather than disempower them through laws.

Our view of mental health and psychological well being has changed considerably over the past century. Once people whose "problems" were considered deviant in society---from mothers who lost children to men who suffered from hallucinations---were often circumscribed to places they wouldn't be seen or heard from; such as the insane asylum where they were often treated with the crudest, most punitive measures such as electroshock therapy and rubber rooms. Although we have evolved greatly as a society since then; it wasn't until very recently that people began to accept the importance of recognizing and admitting psychological issues as normal in most human beings. Accordingly chronic problems such as abuse and addiction were perpetuated blindly for years by former victims who were unaware of how deeply they were affected by their own traumas. Generally, people with less substantial "problems" that were nonetheless affected their lives and choices, were taught to ignore them.

We've come a long way baby. Well, we have come further than we were before. Psychological care is beginning to be recognized as an important component of health, alongside primary health care. Insurance plans are now covering more mental health care services and accompanying holistic services. Hospitals and addiction centers are taking more psychologists and psychiatrists on staff and offering psychological services to their patients.

As times worsen, there is a growing need for psychologists, particularly those specializing in trauma to help with everything from difficulties arising from chronic unemployment to cultural specialists to help refugees displaced from war torn countries to integrate into mainstream society. As the economy collapses and society becomes more complicated, rates of depression, addiction and suicide spike.

As we continue to write laws governing psychologists being seen privately, at drop in centers, in bereavement groups or in institutions; regulating the use and availability of psychotropic drugs such as anti-depressants or anti-psychotics; and regulating those forced into psychological facilities or addiction centers due to their run ins with the law; we must keep human dignity at the forefront of our consciousness. Likewise, we must advocate for legal, insurance and health care regulations that advocate for mental health care to be a top priority, as integral to our lives now as primary health care. This will help psychological problems from escalating to off the charts proportions; people to become more healthy well balanced human beings who are able to take better care of themselves and others on deeper inner levels; and to prevent physical problems and diseases from developing as a result of psychological issues that have gone unchecked.

i) Law and modern technology. A question of balance.

Legal ethics are also important when considering modern technology.

During this decade we are given more---some would argue too much---power through technology, including the power to create or destroy. Some of the more extreme controversial technological advancements include bombs and other devices which could cause nuclear war; the potential for genetic engineering and human cloning.

On the flip side, devices that could greatly aid humanity such as advanced forms of ultrasound, tooth regeneration, advanced water purifiers and pacemakers have been developed or are being developed.

With all this potential, there has to be a series of weights and measures to keep these possibilities in balance. The survival of humanity, the ethics of manipulating humanity or potential for destroying humanity, and new means of creation and helping humanity to adapt to this evolutionary phase of existence must all be brought into light and balance; and considered when creating legal policies designed to regulate society and our lives.

j) Practicing what you preach. Aligning your lifestyle for the good of both self and others.

As you have probably noticed through this book, or in encounters

throughout your own life, once you start talking about these ideas, things start to fall into place and make a lot of sense. Although the qualities we have mentioned throughout this book may seem to run contrary to what we have experienced as children or been taught as adults; they are actually more natural to our way of life as human beings. Most of the work we do as adults entering into a contract with ourselves and spirit to commit towards our own healing and evolution begins with the very difficult process of unlearning the faulty tenants we have been taught, unloading our baggage. Our own personal truths, our authentic selves must be mined for, like precious gems in a watery cave. Only then can we align with our authentic selves and finally with others to work for the greater good.

From a theoretical viewpoint, this all may seem great. We know from experience, it is far trickier to practice what we preach, to bring what we have learned into the world. I suggest that in doing the work to uncover your true selves, unload baggage---powerful things will begin to happen. Everything will fall into place and new opportunities and possibilities will come into your life without your necessarily having to ask for them each and every time. Business partnerships, movements to join, abundance will all come to you; if and only if you may true to yourself and pay attention to the signs, remain ethical to others and commit to discovering your highest calling. From all this, your life purpose will emerge, and the means to put the wheels in motion to make it happen.

I invite you to join me on the next phase of this journey.

Chapter 8

Alternative techniques for walking further down the right path.

a) Tricks of the Trade: Quiet stillness, inspired expression, writing

There are tried and true methods of shifting your life in a general way to move towards achieving anything you desire, including leadership. There are ways to teach yourself to step out of your own way, to leave space in your life for good things to happen, to understand your own heart through quiet stillness and mindfulness.

Quiet stillness like eg meditation has different meanings in different cultural and religious traditions. Essentially mediation is a way of calming the mind and gaining more awareness of self and the world around us.

Sometimes, meditators focus on the breath or another object while sitting in a

relaxed yet fixed position, often with their legs crossed "Indian style" or sitting on a chair with their feet firmly on the floor. Our hands are flat on our thighs, our back is straight but not stiff, our gaze is soft. We pay attention to the way our breath moves throughout our body, simply observing, maintaining our focus. If our mind drifts in and out of focus, becoming momentarily lost in a daydream, an anxiety, a fantasy or thought of the future; we acknowledge what it is doing and gently bring it back.

Trauma and Leadership

We may also mediate on a word, a concept, or something we want to bring more of into our life: for example peace, compassion, healthy love or courage. We might draw our attention to the word itself; that may be our focus and we may repeat it over and over. We may focus our attention on anything that word brings up for us: images, thoughts, memories, examples. This helps us to integrate whatever concept we choose into lives more. If we meditate on these concepts regularly and consistently; amazing things can happen. We may find ourselves feeling more compassion for example for people we relate to in all aspects of our lives; and emotions may be reawakened in us: for example gratitude, love, the impulse to help. We may find ourselves less drawn to self-destructive habits or patterns, less isolated, more peaceful. You can try this exercise with any quality you may want to integrate more into your life or learn more about that you also identify with leadership.

Meditation can also refer to a reflective state. We can be meditative in our everyday life: simply by being mindful of our bodies, of where we are in space and time. For example, if we are doing the dishes, moment to moment, we can bring our attention to what is going on in the present moment: My hands are in the warm sudsy water. It feels heavy, they are moving the sponge over the pan. The sponge is rough on my fingertips. I am scrubbing, It is nighttime. The sky is dark and filled with stars. Simply bringing our attention to the present moment, can restore peace to us, give us more spaciousness inside our minds and bodies to breathe and to be.

Any of these exercises can help us in any leadership role we can take on, and are particularly well suited to preparing us for or being at peace with major changes in our lives. They can help us to trust ourselves, to remain true to our own nature, to be present and effective and loving in our dealings with others.

Inspired expression. Art and Writing. Creation can help us in so many ways. Human beings are creative by nature, but unfortunately, many cultural and societal norms quash the creative impulses in us after we are children. Giving yourself license to create; in whatever magical, frivolous, passionate, wild, quiet way you desire, is one of the greatest gifts you can give yourself in this world. Being creative helps us to connect with that sense of childlike wonder that may have been dormant in us, to play, to contemplate and process, to understand our own hearts and souls in ways that we never knew possible, to see beauty in and feel connection to the mysterious pulsating world around us.

You probably noticed by now that many of the exercises in this book are grounded in writing. This is the language that provides the surest way to sift through all the internal psychological junk we have built up in ourselves and follow a path to our hearts. It is a good way to understand who we are, what we think, how we believe, what we desire. The same is possible through any form of artistic expression: collaging, painting, drawing, singing, music dance. Do whatever it is that you feel drawn to do. You will be amazed at how restorative it can be. It is from this well rested place that we find the true passion. Leadership has deep roots in creation.

Trauma and Leadership

Getting committed. Embodying your answers.

Now that you have made the decision to follow your role as a leader, have

identified your leadership qualities (both those that are most natural to you and those that you may choose to work on), now that you have set your intention, made room in your heart through quiet stillness and mindfulness and processed and played with your intention through writing or art; it is time to commit. Make a commitment to work with your leadership for x amount of time. Find the answers to the questions that arise through this work by feeling; by trusting your intuition and your heart.

Systemic leadership.

The modern world, the transitional world requires that people learn new skills in

systemic and critical thinking to solve problems and join forces with others to find collective solutions to problems. This is effective towards improving issues facing us in the business world, environment or politics. It helps us to coordinate and process our knowledge and awareness of what is still a complex, fragmented society drawn largely towards compartmentalization. It helps us to devise contemporary, multi-level, informed solutions to problems. For example: if we were taking a look at homelessness, we might look at the complex array of social, political, and economic factors that cause people to become homeless; the demographics of the homeless population; potential solutions to homelessness in an isolated area or on a grand scale; and advocates and pioneers of housing alternatives in the homeless community or community of those serving homeless people. In a sense systemic and critical thinking skills are a counterweight to what may seem as an overly esoteric, spiritual or psychological viewpoint---which may seem to some to lack substance out of context. Systemic leadership helps us to follow our heart; and to use all our faculties to make sense of it and use it to the best of our ability: within the context of a broader perspective of society and world. It helps us to set our intentions in a larger context to really work with them.

Just as systemic leadership can help us bring our hearts out to the world; it can   help restore our trust in ourselves after being set adrift due to turmoil brought on by the complexities of the world. It can help us to find our inspiration, heart center, healthy love or intuitive response to a system that has always seemed to have us in its clutches: for example finances, work or large scale economic changes brought on perhaps by the global economy. It can help us to recover from this and give us deeper insight into our skills, talents, clarify a purpose; and design our own map for our careers or other aspects of our lives in the future.

One size doesn't always fit all. Trying new roles through clubs, volunteer work, linking up with a movement, mentor or group.

When all else fails, be brave. Try things you are scared to try. Do things you are scared to do, step outside your comfort zone.

One of the simplest ways to try out new leadership roles is to experiment. You may choose a new group or club to join. You may volunteer to help feed homeless men at a soup kitchen, to stand on line at a soup kitchen yourself to feel what it is like. You may choose to be an advocate for an abused child in court, or be a vigil volunteer and sit by the bedside of people who are in hospice care while they are dying. You may link up with a movement of others working for causes that you have always felt strongly about, or that you have just started to feel impacted by due to the current political climate. You may serve as a mentor, using your skills to teach others. You may attempt to learn a new skill yourself: apprentice for someone who fills out visa applications for recent immigrants, apprentice with a glassmaker or painter to learn new skills to teach to others in the future.

b) Integration. The art of birthing in the unknown and acting upon being.

When we step into the unknown, we are acting with courage.

Human beings are most vulnerable when they step into the unknown for here they find their most authentic, most vulnerable selves; and their purpose.

We step into this realm whenever we are trying something new. When we are trying something new to share with, contribute towards or organize others; we start by stepping into this realm, to align with our most authentic selves. Operating from this place may seem terrifying at first, but little by little it becomes habitual, second nature. As we operate from this realm, we become stronger and more effective leaders for ourselves and for others.

1) Why leadership is crucial for the survival of the planet

a) Recognizing how the problems of the world (genocide, violence, hunger etc) are brought to our doorstep through mass media, increased travel ease, internet and other communication systems.

b) How humans are hard wired to help others.

Recognizing how the problems of the world relate to you. Genuine concern and compassion vs "faking it."

c) The role our compassion plays in shifting world paradigm, aligning with human evolution and a new wave of consciousness.

d) Sustaining interconnectivity for sustainability

How humans are hard wired to help others

The Dalai Lama suggests that human beings are gentle and compassionate by nature---that they are altruistic by default. In other words, when we see an opportunity to help others, we take it. More than this, helping others, helps us.

He also suggests how compassion is a key component of communication. When humans are able to suspend judgement and put their preconceived notions aside, when they are able to work hard to put themselves in another shoes---they can understand and empathize and feel compassion towards that person. This helps us to establish crucial bonds; and with it, a greater sense of belonging to the human race. (The Dalia Lama and Howard Cutler, The Art of Happiness)

Leaders like Deepak Chopra note the reciprocal relationship between giving and receiving, how what you put out into the world comes back to you. This could be due to philosophical/psychological/moral/societal characteristics we learned growing up—and also because there is a scientific universal flow that insures the more we have, the more we give, the more that is given back to us. He coins the term "poverty consciousness" which relates to those who live their lives guided namely by a fear of scarcity--- that there are a limited amount of resources out there and people must fight for them. (Chopra)

The truth is the world is filled with abundance. There is more than enough to go around. When we are greedy we actually squander those resources, we knock things off balance. This is evident in problems of the world ranging from environmental instability and global warming to famine and war perpetuated due to the fear/loss/squandering of resources around the world,

Chapter 9:
Resilience. What is Resilience?

a) How to tap into your hidden strengths to recognize and resist oppression of self and others.

Resilience is the ability to bounce back from tough situations; to rise

up from the ashes of trauma, failure, grief and catastrophe. You may have noticed in your life that there are some people who are more skilled at bouncing back than others; or that everybody has their own time line for bouncing back.

There may be a biological component that makes some people more apt to be resilient than others.

A group of researchers from the Naval Health Research Center and the University of California hypothesized that differences in brain chemistry allow specific individuals to be resilient to stressful situations. Their case study involved Marines, Navy Seals and adventure racers. They concluded that the people who bounce back from stress in order to perform extraordinarily well under duress may have more finely honed interoception---which is informed by an area of the brain called the insula, which has an important job in emotional processing and self-awareness. These processes contributes to the ability to make confident, often split second decisions after suffering adversity and/or in extreme situations.

In these cases it was found that the control groups read certain characteristics, such as the ability to quickly read expressions and emotions on people's faces, and a shorter term reaction and response to stress---and thus had a higher degree of introceptive awareness and more capacity for resilience.

There are those that argue that resilience is not rooted in biology or chemistry; but rather that people's personalities (including their tendency to be optimistic or cheerful), upbringing (learned resilience), support systems, reality, attitude towards change, confidence levels, the ability to recognize triggers and regulate your emotions and physical health all play a role in determining how much resilience a person will have. Likewise, resilience is a quality that can be cultivated and learned.

It is not necessary to determine whether nature or nurture contribute to resilience. It is enough to note that it is a quality that can be tapped into, cultivated or learned.

One of the ways that we can learn resilience is to understand ways in which it may have occurred in our lives in the past.

Try this exercise.

    a) On a piece of paper in your journal make a list of all the times in your life when you struggled, perhaps with things that you do not struggle with anymore, perhaps with things that occur with less frequency than they used to. There are wide ranging possibilities for topics that may have been part of your struggle and may include things such as poverty/financial struggles, physical health problems, bad relationships, addictions (to drugs, alcohol, food, bad partners, exercise, sex), grief over the death of a loved one, depression, suicidal thoughts, trauma, abuse, historical trauma such as slavery, multi-generational trauma such as sexual abuse. Do not be shy and do not get bogged down with the details. Simply list them.

b) Pick one or two of these struggles. Write a few lines, again not getting bogged down with the details, about what occurred during this time period. It may be helpful to use a journalists trick and write a line for each of the 5 W's—who what where when why. Take a deep breath and think about when your reaction to that struggle changed, and when or how you were able to get past it. Right down methods of resilience you might have tapped into to "bounce back." These may include anything. Some common methods of "bouncing back" include: patience, support from loved ones, going through the motions, therapy/understanding what happened, art, becoming more emotionally aware of habitual responses to trauma/what is really driving you.

c) You may not know what caused you to bounce back. It may have simply happened. Write down a line or two about what made you aware of the fact that you bounced back from the struggle, and/or if you haven't been able to bounce back---what strengths you know you have inside you that may cause you to bounce back. These might be simpler than you expect. I often knew many of my deepest struggles had passed, or changed form, when I was out in nature and could again appreciate a beautiful sunset, or when a seal in the ocean reminded me that things were going to be okay. I've realized I'd rounded the hump in depressions in my life when food tasted good to me again, or I woke up and wanted to get up and get dressed and go to work, or when my child kissed me on the cheek and caused me to laugh.

d) Make a note of all the qualities, strengths and talents that you have that you might tap into after a tough situation to bounce back. Are there qualities you don't necessarily have but could cultivate to help you through tough times? For instance: do you feel like you might benefit from more patience, optimism, or compassion? Take this one step further and decide how you might hone these qualities through quiet stilness, art, walks in nature, volunteer work. Do you feel you need to learn more about your emotional reactions to tough situations, that being more emotionally intelligent or aware could help you to weather storms better? How can you achieve this? Would you benefit from traditional therapy, hypnosis, constellation work?

Recognizing ways we have behaved in the face of oppression in the past. Looking at our strengths, weaknesses and habits with honesty.

Being honest with ourselves is the first step in honing our resiliency. Let's be real, when bad things happen, we do not always behave as we would have behaved if we were paying better attention. We do not always emulate the Buddha. We are human, and our human junk tends to get in the way. This of course is how we experience life best, where we learn the most and how we grow and change. When it is happening, it can also really be painful.

1) Write down one struggle that "held you" for a while, or that you still have some unresolved emotions about? What emotions were

triggered while this was happening, what vulnerabilities were exposed, what are some of the ways you reacted that you wish you could change, what are some of the behaviors you exhibited you may feel some embarrassment about. Write these all down.

2) Rip the pages from your book. Read them aloud. Acknowledge they are yours. Own them. Now take them to your sink or to a spot out in nature and burn them, let them go.

3) Make a similar list about the ways you reacted favorably to a struggle that you are proud of. What are some of the qualities you brought to the table that allowed you to process what was going on, get out of your funk, tap into your talents, help others and demonstrate your strengths to others.

4) Rip the pages from your book. Read them aloud. Acknowledge they are yours. Own them. Now take them to your sink or to a spot out in nature and burn them, let them go.

5) How did this exercise affect you? Did it help you acknowledge that our behaviors and emotions and reactions to tough times are ephemeral, neither right nor wrong? Try and keep this exercise in mind on the next phase of your journey.

b) Sharing this strength with others. Setting good examples.

We can look towards others as role models who display extraordinary evidence of resiliency under the most oppressive circumstances. For example, many former Holocaust prisoners have gone on to become famous. Eric Vogel, the infamous jazz trumpeter escaped Nazis while he was being transferred to Dachau Concentration camp. Rose Warfman was a woman who was a member of the French Resistance, helping Jews to resist capture. Even though she was caught, imprisoned in Auschwitz where an operation was performed on her with no anesthesia, she survived and went on to continue to fight for the rights of others.

Everywhere there is oppression, there is resilience. Sojourner Truth was an African American woman who escaped from slavery with her infant daughter in 1826. She had to fight for custody of one of her sons. Her resilience against oppression and a legal system that did not favor former slaves; she stood her ground and became the first black woman to win a court case against a white man; and continued to fight for abolition and women's rights throughout her life. Her speech "Aint I a Woman," spoke volumes on gender inequality and is widely quoted from today.

Iqbal Masih was a Pakistani boy who spent much of his childhood chained to a loom and forced to weave carpets. His early trauma did not break him. He later escaped and went on to become a spokesperson and advocate against child labor in the developing world. His life was constantly in danger, and he was eventually assassinated. Still, he was resilient enough to be brave and bring awareness to child labor violations all over the world.

There are those who fight quietly against injustices they witness happening to people around them, and continue to be resilient to the roadblocks they face. Margaret Sanger spearheaded efforts to bring birth control to the United States and was a key player in establishing the Planned Parenthood Foundation of America. She faced numerous setbacks including more than 9 arrests, the shutting down of each of her clinics, and the passage of the Comstock Law, which lead to the ban on importing or mailing contraception devices. Still, she soldiered on.

There are scores of people who were imprisoned for their resistance to political oppression who became resilient throughout their prison terms to stick to their belief systems or to lead resistance movements and inspire others. Benazir Bhutto was President of Pakistan, twice, and the first woman to lead a Muslim state. She was arrested repeatedly, and spent three years in solitary confinement. Mordechai Vanunu was a former Israeli nuclear technician, whose objection to weapons of mass destruction lead him to blow the whistle on the state of Israel for developing nuclear weapons. He was in prison for 18 years, eleven of which he spent in solitary confinement.

There are others who serve by example, quietly resisting the turmoil around them. There are also countless stories of people who were born in squalor; in favelas or ghettos where there was little hope, and went on to become leaders.

Carolina Maria De Jesus, wrote a book about the conditions in which she lived in a favela on scraps of paper she recovered from the trash.

Trauma and Leadership

Richard Wright and Ralph Ellison were born in inner city neighborhoods where gang activity, drugs and poverty defined the lives of most of their peers, but went on to become famous authors.

Maya Angelo wrote autobiographies about the sexual abuse and prostitution she grew up around. She became the first poet and first woman to speak at presidential inaugurations; including the inauguration of Bill Clinton and Barak Obama.

Exercises

1) Is there a time your resilience in life served as an example for others? Write what you can about that time and how others let you know it affected them.
2) Who are your personal resilience role models?

Lessons from the masters; Qualities that define resilience and how to cultivate them.

As discussed earlier, there are certain characteristics that define

People who are more apt to bounce back. These include overall optimism and a positive attitude, self-awareness and the ability to admit and learn from failure.

Following are some qualities you may cultivate that may help you to be more resilient after tough experiences.

Compassion: People who have suffered who can recognize similar brands of suffering in others and/or have an impulse to help others are more likely to be more resilient.

Being mindful of what is: There is the old AA.NA Adage: God grant me the strength to accept the things I cannot change, to change the things I can't accept and the wisdom to know the difference. This is a quote to keep in mind when dealing with tough times, and trying to bounce back. People who are able to accept and come to terms with what is bad, to admit it and even perhaps to view it as part of your life experience and/or an opportunity for growth and change---tend to be more resilient. Those who can recognize a situation or difficult experience, and see it in more than one context are more apt to be resilient. For example: instead of immediately seeing a traumatic experience, a car accident, as something that happens only to the driver and is going to ruin his/her life; the driver might recognize they need to breathe, and acknowledge all the other people involved. It didn't only happen to me. What do I do now? One step at a time.

The ability to embrace change: Those who are mindful of the way life can change on the flip of a dime, those who may recognize they are uncomfortable when change is occurring, but are open to embracing it anyway, are skilled at resilience. They may be more apt to see difficult situations as transmutable, and change as something they trust in even if it is a tough hill to climb.

Expression: People who are able to find an outlet to transform their emotions and experiences---such as art, writing, or gardening—are able to cultivate their own unique brands of resiliency.

Cultivating self-confidence and faith in your gifts: People who have a strong core belief in who they are and what they can do, tend to be able to respond to crisis in a more positive way.

Optimism : Those who can remain hopeful about the future, even in the face of the most difficult circumstances, are better adaptable.

Finding a Support System: Those who have friends, family, colleagues, therapists, or groups to rely on when they are experiencing tough times;l tend to bounce back better. Having people to share your feelings with, guide you through wisdom and feedback, and to advise and/or witness what you are going through can help.

Taking care: It is important to exercise self care, to be kind and nurturing to yourself when times are hard. Take care of yourself as if you were your own child or a sick relative: make sure you eat right, exercise, sleep and do things you truly love.

Become a good decision maker: People who are able to develop problem solving abilities, who are able to make tough choices and then live with them and let them go, who see more than one solution to a problem, are apt to bounce pack from tough times.

Be Active: Recognizing your problems will not go away on their own, is your first step to dealing with them and being able to ultimately move on.

Mindfulness: Another good tool for developing resilience is mindfulness training. This can be achieved through formal quiet stilness (x minutes of sitting and/or walking quiet stilness per day) focused on breathing or a particular quality you would like to cultivate in your life such as compassion, forgiveness, confidence, truly love. You may also practice mindfulness in your life simply by being mindful, by checking in with your thoughts or physicality every once in a while (in this moment I am washing the dishes, the hot water feels good on my hands, the plate is smooth, I am grateful for the food I just ate, for the ability to clean). You might check in with yourself about a particular problem or specific emotion arises; being cognizant that it is occurring—How interesting, I am really angry right now. What triggered this anger? What is the real cause of this emotion?

2) Why leadership is crucial for the survival of the planet

e) Recognizing how the problems of the world (genocide, violence, hunger etc) are brought to our doorstep through mass media, increased travel ease, internet and other communication systems.

f) How humans are hard wired to help others.

g) Recognizing how the problems of the world relate to you. Genuine concern and compassion vs "faking it."

h) The role our compassion plays in shifting world paradigm, aligning with human evolution and a new wave of consciousness.

i) Sustaining interconnectivity for sustainability

j) Recognizing how the problems of the world (genocide, violence, hunger etc) are brought to our doorstep through mass media, increased travel ease, internet and other communication systems.

One important way the world mass consciousness is changing is that

we are more interconnected through the Internet, cell phones, television and radio. People are being introduced to the problems facing people in other parts of the world and their own countries often immediately after or as they are happening. Additionally increased access to travel; and more and more immigrants and refugees moving to new locations; people are becoming more cognizant of the very different cultural/economic/religious/social realities facing groups of people who we come into increased contact with. Likewise, as we become close to other groups of people; as they become members of our neighborhoods, workplaces, school yards; as they become our friends and our families---we start to both appreciate and become influenced by their differentness; and become aware that as members of the human race, there aren't really that many differences.

When this happens, when we become cognizant of others problems in a relatable way, we care. Empathy naturally arises, and with it, often, compassion, the drive to do more to help. We become quite literally interconnected. Then, even the sufferings of others in other parts of the world aren't so foreign to us. If we know somebody intimately who has been through a hurricane in Louisiana where they lost much of their family, it may remind us of a tragic death of a family member to a natural disaster in our own family. If we see on television people in India who have fallen victim to a landslide, if we see the ways they are suffering because the people they love are dying—it then becomes more relatable to us.

Of course, there is also a school of thought that access to so much information and such intense awareness of the ways people are suffering in so many different ways all over the world---could overwhelm us and make us numb to what is happening. This is also a possibility. It may also be the case that we need to experience suffering collectively just as we do as individuals, in order to heal.

c) B) How humans are hard wired to help others

Have you ever helped anyone without any ulterior motives, without wondering what was in it for you, and then found, that in the process of helping them and afterwards you felt restored, as if you yourself had returned to some innate place in yourself where things felt right.

In societies with a survival of the fittest, dog eat dog, get them until they get you, fake it until you make it mentality; people are often diverted from their natural impulse to help. We are often instructed to tune out the suffering of others, and that this is necessary in order to ensure our own self-preservation. We pass some friends off to therapists, bide our time and wait until others hit "rock bottom" before we feel inclined to help. Often, when we help someone, we feel compelled to hide that fact. We may drop change into a homeless person's jar, then quickly run away to disguise the fact in case anyone judges us.

In reality, human beings are genetically hard wired to help each other. Social scientists and neuroscientists have been studying or the past decade or so the way primitive parts of the brain---most remarkably those reward centers that light up as a response to excitement over food or sex—were also activated when subjects were given the ability to help others. In this sense being altruistic is actually natural to us and brings us pleasure; in contrast to thoughts that helping others is a hardship and something people do begrudgingly because they are aligning with their morals. Moreover, we are hard wired to help because we are hard wired to empathize with the suffering of others.

It is hard to get back to that place in ourselves and our society where helping is a natural and integral part of our daily lives, particularly if we work in corporate environments and have busy lives. Yet doing so can be just as, if not more, beneficial to you than it is to the ones you are helping. Helping can align us with our basic nature as human beings, and as a society.

The role our compassion plays in shifting world paradigm, aligning with human evolution and a new wave of consciousness.

I believe this kind of consciousness will naturally arise in us as

the world and human beings evolve. We will be drawn more to understand and empathize with each other; and with the sufferings of the world at large. Accordingly, we will be driven to help. With more people helping, our consciousness will evolve, and there will be more opportunities to help integral to daily life. Helping will become more natural and ordinary, and we will be better off for it.

d) Sustaining interconnectivity for sustainability

When we are interconnected we are stronger. As a unified force we are more apt to come up with solutions to societal and global problems; and less likely to self destruct. Just as we are hard wired to help, we are also hard wired, on a cellular level to interact with other human beings. As we break free from the isolation and hermitage many of us have adapted in this society; we will be in better shape, our lives will be more conducive to healing ourselves and the planet.

3) Resilience.

a) How to tap into your hidden strengths to recognize and resist oppression of self and others.

b) Quiet listening skills for befriending the other.

c) Sharing this strength with others.

d) Recognizing ways we have behaved in the face of oppression in the past. Looking at our strengths, weaknesses and habits with honesty.

e) How to "own" our reactions to oppression as a means of transformation in the future.

Resilience.

What is Resilience?

a) How to tap into your hidden strengths to recognize and resist oppression of self and others.

Resilience is the ability to bounce back from tough situations; to rise

up from the ashes of trauma, failure, grief and catastrophe. You may have noticed in your life that there are some people who are more skilled at bouncing back than others; or that everybody has their own time line for bouncing back.

There may be a biological component that makes some people more apt to be resilient than others.

A group of researchers from the Naval Health Research Center and the University of California hypothesized that differences in brain chemistry allow specific individuals to be resilient to stressful situations. Their case study involved Marines, Navy Seals and adventure racers. They concluded that the people who bounce back from stress in order to perform extraordinarily well under duress may have more finely honed interoception---which is informed by an area of the brain called the insula, which has an important job in emotional processing and self-awareness. These processes contributes to the ability to make confident, often split second decisions after suffering adversity and/or in extreme situations.

In these cases it was found that the control groups read certain characteristics, such as the ability to quickly read expressions and emotions on people's faces, and a shorter term reaction and response to stress---and thus had a higher degree of introceptive awareness and more capacity for resilience.

Trauma and Leadership

There are those that argue that resilience is not rooted in biology or chemistry; but rather that people's personalities (including their tendency to be optimistic or cheerful), upbringing (learned resilience), support systems, reality, attitude towards change, confidence levels, the ability to recognize triggers and regulate your emotions and physical health all play a role in determining how much resilience a person will have. Likewise, resilience is a quality that can be cultivated and learned.

It is not necessary to determine whether nature or nurture contribute to resilience. It is enough to note that it is a quality that can be tapped into, cultivated or learned.

One of the ways that we can learn resilience is to understand ways in which it may have occurred in our lives in the past.

Try this exercise.

    a) On a piece of paper in your journal make a list of all the times in your life when you struggled, perhaps with things that you do not struggle with anymore, perhaps with things that occur with less frequency than they used to. There are wide ranging possibilities for topics that may have been part of your struggle and may include things such as poverty/financial struggles, physical health problems, bad relationships, addictions (to drugs, alcohol, food, bad partners, exercise, sex), grief over the death of a loved one, depression, suicidal thoughts, trauma, abuse, historical trauma such as slavery, multi-generational trauma such as sexual abuse. Do not be shy and do not get bogged down with the details. Simply list them.

b) Pick one or two of these struggles. Write a few lines, again not getting bogged down with the details, about what occurred during this time period. It may be helpful to use a journalists trick and write a line for each of the 5 W's—who what where when why. Take a deep breath and think about when your reaction to that struggle changed, and when or how you were able to get past it. Right down methods of resilience you might have tapped into to "bounce back." These may include anything. Some common methods of "bouncing back" include: patience, support from loved ones, going through the motions, therapy/understanding what happened, art, becoming more emotionally aware of habitual responses to trauma/what is really driving you.

c) You may not know what caused you to bounce back. It may have simply happened. Write down a line or two about what made you aware of the fact that you bounced back from the struggle, and/or if you haven't been able to bounce back--- what strengths you know you have inside you that may cause you to bounce back. These might be simpler than you expect. I often knew many of my deepest struggles had passed, or changed form, when I was out in nature and could again appreciate a beautiful sunset, or when a seal in the ocean reminded me that things were going to be okay. I've realized I'd rounded the hump in depressions in my life when food tasted good to me again, or I woke up and wanted to get up and get dressed and go to work, or when my child kissed me on the cheek and caused me to laugh.

d) Make a note of all the qualities, strengths and talents that you have that you might tap into after a tough situation to bounce back. Are there qualities you don't necessarily have but could cultivate to help you through tough times? For instance: do you feel like you might benefit from more patience, optimism, or compassion? Take this one step further and decide how you might hone these qualities through quiet stilness, art, walks in nature, volunteer work. Do you feel you need to learn more about your emotional reactions to tough situations, that being more emotionally intelligent or aware could help you to weather storms better? How can you achieve this? Would you benefit from traditional therapy, hypnosis, constellation work?

Recognizing ways we have behaved in the face of oppression in the past. Looking at our strengths, weaknesses and habits with honesty.

Being honest with ourselves is the first step in honing our resiliency. Let's be real, when bad things happen, we do not always behave as we would have behaved if we were paying better attention. We do not always emulate the Buddha. We are human, and our human junk tends to get in the way. This of course is how we experience life best, where we learn the most and how we grow and change. When it is happening, it can also really be painful.

a) Write down one struggle that "held you" for a while, or that you still have some unresolved emotions about? What emotions were triggered while this was happening, what vulnerabilities were exposed, what are some of the ways you reacted that you wish you could change, what are some of the behaviors you exhibited you may feel some embarrassment about. Write these all down.

b) Rip the pages from your book. Read them aloud. Acknowledge they are yours. Own them. Now take them to your sink or to a spot out in nature and burn them, let them go.

c) Make a similar list about the ways you reacted favorably to a struggle that you are proud of. What are some of the qualities you brought to the table that allowed you to process what was going on, get out of your funk, tap into your talents, help others and demonstrate your strengths to others.

d) Rip the pages from your book. Read them aloud. Acknowledge they are yours. Own them. Now take them to your sink or to a spot out in nature and burn them, let them go.

e) How did this exercise affect you? Did it help you acknowledge that our behaviors and emotions and reactions to tough times are ephemeral, neither right nor wrong? Try and keep this exercise in mind on the next phase of your journey.

e) Sharing this strength with others. Setting good examples.

We can look towards others as role models who display extraordinary evidence of resiliency under the most oppressive circumstances. For example, many former Holocaust prisoners have gone on to become famous. Eric Vogel, the infamous jazz trumpeter escaped Nazis while he was being transferred to Dachau Concentration camp. Rose Warfman was a woman who was a member of the French Resistance, helping Jews to resist capture. Even though she was caught, imprisoned in Auschwitz where an operation was performed on her with no anesthesia, she survived and went on to continue to fight for the rights of others.

Everywhere there is oppression, there is resilience. Sojourner Truth was an African American woman who escaped from slavery with her infant daughter in 1826. She had to fight for custody of one of her sons. Her resilience against oppression and a legal system that did not favor former slaves; she stood her ground and became the first black woman to win a court case against a white man; and continued to fight for abolition and women's rights throughout her life. Her speech "Aint I a Woman," spoke volumes on gender inequality and is widely quoted from today.

Iqbal Masih was a Pakistani boy who spent much of his childhood chained to a loom and forced to weave carpets. His early trauma did not break him. He later escaped and went on to become a spokesperson and advocate against child labor in the developing world. His life was constantly in danger, and he was eventually assassinated. Still, he was resilient enough to be brave and bring awareness to child labor violations all over the world.

There are those who fight quietly against injustices they witness happening to people around them, and continue to be resilient to the roadblocks they face. Margaret Sanger spearheaded efforts to bring birth control to the United States and was a key player in establishing the Planned Parenthood Foundation of America. She faced numerous setbacks including more than 9 arrests, the shutting down of each of her clinics, and the passage of the Comstock Law, which lead to the ban on importing or mailing contraception devices. Still, she soldiered on.

There are scores of people who were imprisoned for their resistance to political oppression who became resilient throughout their prison terms to stick to their belief systems or to lead resistance movements and inspire others. Benazir Bhutto was President of Pakistan, twice, and the first woman to lead a Muslim state. She was arrested repeatedly, and spent three years in solitary confinement. Mordechai Vanunu was a former Israeli nuclear technician, whose objection to weapons of mass destruction lead him to blow the whistle on the state of Israel for developing nuclear weapons. He was in prison for 18 years, eleven of which he spent in solitary confinement.

There are others who serve by example, quietly resisting the turmoil around them. There are also countless stories of people who were born in squalor; in favelas or ghettos where there was little hope, and went on to become leaders.

Carolina Maria De Jesus, wrote a book about the conditions in which she lived in a favela on scraps of paper she recovered from the trash. Richard Wright and Ralph Ellison were born in inner city neighborhoods where gang activity, drugs and poverty defined the lives of most of their peers, but went on to become famous authors.

Maya Angelo wrote autobiographies about the sexual abuse and prostitution she grew up around. She became the first poet and first woman to speak at presidential inaugurations; including the inauguration of Bill Clinton and Barak Obama.

Exercises

1) Is there a time your resilience in life served as an example for others? Write what you can about that time and how others let you know it affected them.
2) Who are your personal resilience role models?

Lessons from the masters; Qualities that define resilience and how to cultivate them.

As discussed earlier, there are certain characteristics that define

People who are more apt to bounce back. These include overall optimism and a positive attitude, self-awareness and the ability to admit and learn from failure.

Following are some qualities you may cultivate that may help you to be more resilient after tough experiences.

Compassion: People who have suffered who can recognize similar brands of suffering in others and/or have an impulse to help others are more likely to be more resilient.

Being mindful of what is: There is the old AA.NA Adage: God grant me the strength to accept the things I cannot change, to change the things I can't accept and the wisdom to know the difference. This is a quote to keep in mind when dealing with tough times, and trying to bounce back.

People who are able to accept and come to terms with what is bad, to admit it and even perhaps to view it as part of your life experience and/or an opportunity for growth and change---tend to be more resilient. Those who can recognize a situation or difficult experience, and see it in more than one context are more apt to be resilient. For example: instead of immediately seeing a traumatic experience, a car accident, as something that happens only to the driver and is going to ruin his/her life; the driver might recognize they need to breathe, and acknowledge all the other people involved. It didn't only happen to me. What do I do now? One step at a time.

The ability to embrace change: Those who are mindful of the way life can change on the flip of a dime, those who may recognize they are uncomfortable when change is occurring, but are open to embracing it anyway, are skilled at resilience. They may be more apt to see difficult situations as transmutable, and change as something they trust in even if it is a tough hill to climb.

Expression: People who are able to find an outlet to transform their emotions and experiences---such as art, writing, or gardening—are able to cultivate their own unique brands of resiliency.

Cultivating self-confidence and faith in your gifts: People who have a strong core belief in who they are and what they can do, tend to be able to respond to crisis in a more positive way.

Optimism : Those who can remain hopeful about the future, even in the face of the most difficult circumstances, are better adaptable.

Finding a Support System: Those who have friends, family, colleagues, therapists, or groups to rely on when they are experiencing tough times;l tend to bounce back better. Having people to share your feelings with, guide you through wisdom and feedback, and to advise and/or witness what you are going through can help.

Taking care: It is important to exercise self care, to be kind and nurturing to yourself when times are hard. Take care of yourself as if you were your own child or a sick relative: make sure you eat right, exercise, sleep and do things you truly love.

Become a good decision maker: People who are able to develop problem solving abilities, who are able to make tough choices and then live with them and let them go, who see more than one solution to a problem, are apt to bounce pack from tough times.

Be Active: Recognizing your problems will not go away on their own, is your first step to dealing with them and being able to ultimately movve on.

Mindfulness: Another good tool for developing resilience is mindfulness training. This can be achieved through formal quiet stilness (x minutes of sitting and/or walking quiet stilness per day) focused on breathing or a particular quality you would like to cultivate in your life such as compassion, forgiveness, confidence, healthy love. You may also practice mindfulness in your life simply by being mindful, by checking in with your thoughts or physicality every once in a while (in this moment I am washing the dishes, the hot water feels good on my hands, the plate is smooth, I am grateful for the food I just ate, for the ability to clean). You might check in with yourself about a particular problem or specific emotion arises; being cognizant that it is occurring—How interesting, I am really angry right now. What triggered this anger? What is the real cause of this emotion?

Inner peace for outer peace.

It is true that wars, poverty, and conflict in society reflect the

Internal conflict of individuals, a group of individuals or collective society. While it may seem intuitive to look at a problem---for example hunger in a developing nation---and attempt to fix it by attacking the problem. We may do this. We may persecute a leader for hoarding food sent by development agencies intended for the general population, or for trying to exploit the poor in their country for their own gain; or for stealing crops or indenturing poor farmers. This may very well alleviate the problem for a while. However, it is also important to recognize that the source of the problem does not just lie within that individual leader. The cycle that perpetuated itself within that leader's reign has affected everyone involved. Within time, another leader will be greedy and take his place; more poor farmers will feel insecure about their food supplies. When we deal with a problem, we must also reach out on a deep psychological level to all the people affected, if we ever expect true healing in a society.

Take the example of a genocide. A woman watched her husband and sons being killed in front of her house. She bears deep scars from that trauma. The soldiers who murdered her husbands and sons may also be scarred. The neighbors who watched, the nurse who took care of the woman when she went mad, the woman's other children and their children are all affected. Healing must take place on deep levels for many people before it can be manifested in the world.

These are of course extreme examples and in some cases, it is too late to help people directly involved in situations like these find true peace that will then be manifested in the world. However, each of us bears our own scars from our own traumas, each of us has our own obstacles to finding both healing and inner peace. It is something to work towards. When human beings begin to heal on their own, to find their own inner peace; we can then project and reflect that peace onto the world. It is the direction in which we are headed.

a) There is enough for everyone.

One of the biggest obstacles to peace is the scarcity myth---the belief

That there are not enough resources for everyone, that we have to fight others in order to secure abundance for ourselves. This is where the survival of the fittest mentality that has recently lead to ethical discrepancies and breaches in the business world, particularly on Wall Street where leaders selfishly manipulated other people and abused their own power in order to promote their own self interests. This mentality is also evident on smaller scales: on the way we conduct business and even in our relationships. There is a long standing belief that if you have, others have not.

In reality this is not the natural order of things. We have more than enough resources to go around. There is currently enough grain to feed the world, enough water to quench our thirst, enough energy to power our lives and more than enough love for everyone. Fear based on scarcity mentality is hard to chip away at. In order to advance as human beings we might take a leap of faith, trust that we don't have to step on others to get what we want and that the world will provide for us if we let it.

Cultivating peace and compassion. Cultivating inner peace for clarity, creativity and communication

We have probably all noticed how when we have a problem, we

may feel spiritually, emotionally, creatively blocked. We may be physically unable to complete work. We may find our voice blocked, and have difficulty communicating with others.

We also may have noticed how when that problem is solved, or the pain or suffering associated with it is alleviated, we may find ourselves gain clarity, find ourselves with new creative energies or communicating more freely. Attaining some degree of inner peace opens up our internal channels for energy and power to flow more freely.

Peaceful aligned intentions and visions for success: money, goals, massive prosperity

The interesting phenomenon that occurs when you trust—that the inner resources will provide, that there is more than enough to go around---and when we begin to heal some of our inner turmoil---gifts come to us. We may find ourselves meeting our goals with unexpected ease. We may find money flowing more abundantly in our lives, even when we are not actively seeking it out. We only need to open ourselves up to abundance, to accept it, to have it.

Sometimes, things do not occur quite so seamlessly and automatically. We can set the stage for abundance and prosperity to arrive, by practicing. Here are some exercises you might try.

a) Set a clear intention about how you would like your future to look as it pertains to abundance. You can set more than one. Write them down. For example: I would like to double my income this year. I would like to have enough resources to set up my own adoption agency in the next five years.

b) Pick a time of day or night when you feel most relaxed. Light some candles, burn some sage or incense, dim the lights. Sit or lie in a comfortable position. Say your intention out loud. Breathe in and out four or five times, slow, deep breaths. Wait. Try to envision what your life would be like if that occurred. When you see it, try to really feel it. Let your vision form.

c) Later, in your notebook, set some clear steps you can take to achieve your goals, dreams and visions. Keep it simple. Write whatever comes to you. Pick four or five steps, and set dates when you would like to put your plans into action.

d) In all these exercises, try to maintain alignment with your most peaceful self. "Do what you would do if you felt most secure."

From give-take to give-give.

A wise person once told me the best thing you can do is to help people without expecting anything in return. We are taught the opposite: that we should never give to people if we don't get anything in return. Yet something amazing happens when we shift that paradigm. If we give and give and give; without expecting anything in return: we get a lot more than we could have ever imagined.

Birthing a new renaissance in business.

As the world evolves, we have a chance to recreate ourselves, our

buisness and society according to new paradigms. In the business world, this is already occurring. Calls for more conscientious, ethical and humane business practices have caused old businesses to revamp their codes and new businesses to emerge. Many businesses give a portion of their proceeds to needy people, or encourage or mandate their employees volunteer to help others. At the same time, many businesses are integrating new environmental standards: adhering to building codes that incorporate green building practices: utilizing solar energy, passive lighting systems, sustainable waste water.

There are non profit businesses developed specifically for the purpose of helping others: those geared towards human rights, womens rights, sustainable development, homelessness, immigration or helping victims of child abuse or domestic violence.

As the world evolves, more businesses will emerge incorporating larger groups of people with similar ideologies or from different regions or reflecting different groups needs and best practices. We anticipate some of these businesses will be sourced and motivated by true caring and concern instead of profit margins.

Chapter 10:
Taking our Cues from our childrenPreparing our children to be future leaders.

Parenting and shame

In Brene Brown's Book, I thought it was Just Me, she notes how her interviews with women revealed they often felt judged, criticized or shamed by other women at the way they raised their children. She notes that women are often shamed when it comes to the way they raise their children; where as men are often held to similar forms of scrutiny when it comes to their financial or professional success or non success.

It is tough to be a parent. There are no instruction manuals. Parenting is more challenging in the modern era where parents often must work long hours in addition to their parenting; often simply to put food on the table and make sure their children are fed and they pay the rent. The extended support systems that were once in place through extended family and neighbors; are often replaced with child care facilities and schools. Advice and guidelines for parenting on topics ranging from health to discipline are as likely to come from popular culture---ie, television news and radio programs; as they are from relatives or friends.

With all this uncertainty, it is likely that judgment might occur.

Brown notes that judgment begats judgment---that fear of being judged also causes us to judge others; and that it is nearly impossible to fake being "non judgmental as it appears in our expressions and body language. In order to break this chain, she suggests we be more mindful of what we are thinking, feeling, seeing in regards to parenting; and where our opinions are really rooted. Of course, a good technique for doing this is to go back to our own childhoods, again, to uncover the proverbial skeletons in our closet, to bring to conscious awareness our subconscious motivations, to root them out. This will help us both in interacting and forging communities and trust with other parents; and in our own parenting. Judgment does not necessarily come from the outside; but to the critical voices of our own parents and peers from the past I our heads.

Parenting and Childrens image

Body Image

Brown also notes that parents are apt to pass on shame to their children as it relates to self image, particularly body image and attractiveness for girls, and that this shaming can cripple a child's spirit and self esteem; and is one factor that can prevent them from being strong leaders and role models for others.

Boosting a child's self esteem by helping them to affect positive and realistic body and appearance images; by emphasizing for instance health over weight loss, or inner beauty in conjunction with outer beauty can help them to be less bogged down with narcissitic and neurotic expectations of themselves; and more fully capable, self reliant human beings in the future.

Brown notes that body shaming has far reaching impacts into how we love, live and parent.

Parenting and perfection

When we are parenting, we should always cut ourselves a little slack. Children are difficult, and we are fallible human beings. There is no such thing as perfect parenting. There is no sure fire recipe that will yield a perfect little human being. We all learn from our mistakes---children and parents. Imposing unrealistic expectations on children whether it be for their academic achievements, friendships or behaviors can cause great harm; often having the opposite effect as we desired.

Brown notes that we are particularly apt to attempt to "shame, terrify or judge" our children when they are not living up to our expectations of how we would like to be perceived as parents, when they do not live up to our ideal "parenting image." It would behoove us to attempt not to inflict that on our children, to acknowledge they are independent human beings that do not "belong" to us and do our best to allow them to live out their personalities and roles in the world. We must of course exercise proper judgment when it comes to parenting, set guidelines, restrictions and expectations for behavior and achievements; but it is best to keep these realistic and always acknowledge our child's uniqueness and independence from us.

When things go wrong. Parenting and shame

There is no perfect human being. We learn as humans through our mistakes, through a series of unique trials and tribulations through which we experiment and often "mess up." The ways in which we mess up are not necessarily exclusively disasters; because they teach us and provide us perspective through which we measure other trials and tribulations in our future lives.

Parenting can be toughest when our children are growing and we recognize they are in serious trouble. We may have a child who grows up depressed or with mental health issues. We may have a child who has issues with addiction. We may have a child who winds up in an abusive relationship or being an abuser. We may have a child who has problems with money. We may have a child who robs a bank.

Some of the toughest things to do in this situation are to take a step back and allow our child to mess up. It is tough to resist trying to fix it, blaming ourselves; blaming, shaming or passing judgment on him/her.

Sometimes, tough situations require we exercise tough healthy love---allowing our children to know we are there for them and support them and will be there if they are willing and ready to make a shift in their life; but that we do not necessarily agree with or support their behaviors. Sometimes loving someone requires we take a step back and allow them to come back to themselves.

Sometimes, our children's troubles are the result of our own troubles. This can be hard to see, and to admit. Our aggression or addictions or bad choice in partners may have contributed to our child's depression, violence or addictions. If we know this to be the case, sometimes the bravest thing we can do is to sit down with him/her and admit what we know and what we have done; to apologize if necessary, to offer to help him/her with therapy or other support networks to get help.

Again, with parenting comes great responsibility. Parents in this generation are lucky in a sense because we have come to a place where we are beginning to deal with our own trauma, and to attempt not to pass these traumas down to our children. Our children are growing up with more support to deal with their own trauma; and are less likely to develop habitual reactions to trauma and to have outlets that keep them from being re-traumatized.

Children who have dealt with their trauma or who have less trauma due to the fact that we are aware of our own---are more likely to grow up to be powerful, independent human beings who trust and love others. They are more likely to be affective guides and leaders for others.

Shifting the shame blame paradigm

Brown notes that the influence parents have over their children's ego, self- esteem and sense of potential is boundless. It never goes away. Likewise, if parents have unwittingly and unintentionally used shame, blame, ridicule or judgment as a means of trying to control, motivate or influence behavior; these feelings might stick. More often than not, these are habits we learned from our own parents, that we carried with us on a subconscious level our whole lives. Again, awareness and ownership of our own problems and traumas, can pave the way for our own healing and for our child's healing.

The good news is we are with stuck with our children for the length of their lives and they are stuck with us. We have plenty of time to make things right, to heal our relationships. Brown notes that simply acknowledging some real or perceived wrong we may have done to our child---can have an extraordinary impact. For example simply saying "I understand how my actions made you feel," or "I'm sorry," without overanalyzing or defending can really help.

Resilience in children

The bad news is we can really mess our children up. The good news is we can only mess them up to a point. Children are remarkably resilient. They are also hard wired to be proud of who they are, and to know who they are.

We can also teach our children to increase their resilience, particularly in the face of adversity or hatred, blame or disconnection.

The best thing we can do with parents is to attempt to raise our children with the qualities we cultivate in our own lives. These may include things like empathy and compassion for other being, being of service to others, bravery in the face of adversity or despite fear, and a sense of capability and pride in who they are and what they can accomplish. We can consciously choose to build these qualities in our children. We can teach them to connect with others, to see beauty, to love and to trust. We can teach them that they belong in the world, and that the world is not such a scary place.

Children and connection

One of the biggest problems with today's society is the dissociation and isolation endemic to it. People are taught not to trust each other, to be terrified of the world and the people in it, that there is not enough to go around. Children learn these values from their parents. If we alter our realities so that we are more connected emotionally, intellectually and physically to others---we will also be altering our children's realities. If we teach them to rely on others as well as themselves, to teach others as well as themselves, to help others as well as themselves---we will be paving the way for a shift of consciousness and change in the world.

One of the most important things we can do is to teach our children that they are rooted in the earth, that they belong here, that we all belong. There are unique gifts and roles we all have to be celebrated. The earth is a rich, abundant place with plenty of resources to go around. Our children must be taught how integral those resources are to them, and how fragile; and to respect and defend them. We all have a right to be here. We all have to stand together to be here, to shift the consciousness of the world and to heal ourselves and each other. It is only when we stand together, when we all admit we belong, that this kind of change can occur.

Embrace your child's unique gifts

Artist, introvert, leader,

When it comes to your child's talents, gifts or roles in the world; it is bravest to watch them and let them live out their roles in the world; without trying to impose your own expectations or value system. Support and nurture the talents they have; acknowledge things that are difficult for them not necessarily as weaknesses or obstacles that they have to overcome and do what you can to help them to plug along. Help your child come into his/her own, but allow him/her to lead the way.

This society can often be perceived as self destructive. There are clear cut disciplines and roles that we tend to give more respect to or credence to, such as earning money, being outgoing, talking a lot or loudly or in a convincing fashion, being overly social, being good in business. There are roles we sometimes tend to give less credence to; such as creativity, musicality, introversion, thoughtfulness, what is perceived of as passivity. Things are changing and we may come to realize we need more of these qualities in the future. There are roles children are being groomed for we may not even know exist yet; such as an extraordinary aptitude at technology, making use of other sensory perceptions for example future chefs with a more pronounced ability to taste the nuances of food or future therapists with a more advanced ability to read or empathize with others.

In Susan Cain's book, Quiet, The Power of Introverts in a World That Can't Stop Talking, she implores parents of introverted kids to embrace their childs introverted qualities; and that while parents of introverted children may worry that they will not be accepted in or fit into society; it would behoove them to note that they are probably not going to be able to change their fundamental nature. Likewise, she notes, those introverted qualities may be developed during the course of a child's life and yield good results such as a greater capacity for self-awareness, thoughtfulness, observation skills, the ability to listen to and empathize with others on deeper levels. Although it may be difficult or troublesome for parents to accept these qualities in their children; it would be in their best interest to take a step back and see what the world looks like from the perspective of an introverted or quieter child.

She also suggests that some of the commonalities shier children might exhibit is the need for more down time or quieter environments, breaks between activities, and perhaps slower but deeper levels of conversation. They may need

Whatever your child's gifts are, embrace them. If they are boisterous and active and engaging, but have trouble focusing on a single task try and find a multitude of activities or sports they enjoy to keep them busy. If they like making a mess, introduce them to an art project. If they are orderly and overly organized to the point of seeming a little OCD---give them a recipe book to catalogue.

It is also good to ease them into situations that are uncomfortable to them. If a child is extremely introverted to the point of extreme isolation, it may be good to introduce them to other children and groups. If your

Trauma and Leadership

child is an extrovert to the point of distraction, it may be good to attempt to introduce them

Children and leadership

There are several ways we can prepare children to be the kind of leaders we will ultimately need in the world.

One of the most important things we can do is to change ourselves, to work to break the cycles of trauma that have existed in our lives; and not to pass it on to them. We can try our best not to shame, judge, or criticize them; and if we do to acknowledge this and to acknowledge the ways we make them feel. We can try and embrace their talents and gifts and unique nature; to teach them to be proud of who they are and to love.

We can prepare the ground for the kind of world in which they will grow. This includes introducing them to many different groups of people and teaching them to celebrate cultural, racial and economic diversity. We can teach them how important it is to care for the natural resources of the world. We can serve by example: pointing out where injustices, inequities or violence exists in the world---and the importance of joining movements protesting or resisting these things. We can teach them to trust themselves; to connect with, trust and love each other. We can whenever possible, follow their example, learn about the world they are going to inherit through their actions and gifts unique to them and their generation.

Chapter 10
Traumapsychology, birth trauma/trauma resulting from childbirth, early childhood trauma and how to apply leadership principles for healing.

a) Conscious vs unconscious realities

Many of us spend a lot of our lives believing we are living fully consciously aware; yet being driven, triggered, motivated by habits, patterns and beliefs that function on a largely subconscious or even unconscious level. We live in societies where many people live their lives without full conscious awareness. Accordingly, collectively, we are also functioning largely on a less conscious level.

Factors that may keep us living without full conscious awareness as individuals include trauma---such as abuse, racism, birth trauma---addictions, physical injuries, mental health issues or some forms of oppression. Factors that may keep us living without full conscious awareness collectively include violence (genocide, war, rape), political oppression (lack of freedom of speech, restrictive movement/curfews), class restrictions (such as a caste system) or unequal distribution of wealth (gap between the haves and have not's). We may also be scarred on subconscious levels due to shifting collective perception, for example when the media inundates us with images of terrorism or disease (such as ISIS and Ebola) and people become scared on a deep, primal level; believing they are not safe.

As Dr. Franz Ruppert notes, when we are consciously aware, we function better both in our own lives and in society. We tend to have better communication with others, be more effective in problem solving and our planning.

In order to become more consciously aware; we have to first become aware of what is driving us subconsciously---both as individuals and a society. When there is a trauma, we might seek to identify the events that caused this trauma and recover the memories associated with it.

b) Human beings as unique---Trauma as scarring

The human psyche is remarkable. It transforms reality as it is, to subjective reality for the bearer. It is our main means of perceiving, empathizing, thinking, remembering, dreaming and feeling. It is our multidimensional, but it is also selective. This means there is the potential for perception to be limited, to provide us with faulty information, to be self-destructive.

Just as whales have unique tail prints that they leave in their wake when they slap their tails down on the water, that scientists can use to identify them; so does each individual have a unique life print stemming from their unique psychic structures stemming from their life experiences, their personal history and their genetics

Our collective realities can also have unique characteristics due to the time and countries we are living in, due to our countries history, due to our socioeconomic or political realities; or the greater socioeconomic or political standing of our country.

As we have noted elsewhere in the book trauma can occur on many levels which can scar us.

c) Multi-generational trauma. Violence.

Abuse or other forms of violence can be passed down through generations—in obvious forms such as physical abuse, verbal abuse, rape on incest; or in more subtle ways such as prejudice and hatred.

Often the roots of this trauma are passed down from our ancestors, or several generations of our families without then being consciously aware of them. They may have started on a political level; for example an ancestor who was persecuted during a war or was forced to be a persecutor—committing violence on others---may take his anger or grief out on his family without recognizing what he is doing. A woman may grieve over the loss of a parent and become depressed (anger towards inwards) and not realize she is neglecting her children. Children who are affected by this neglect may wind up being affected in various manners such as contending with addictions, aggression, suicidality; or neglecting or being angry towards their own children.

Multi-generational trauma can also take place on a collective level as well. Traumas passed down over generations can effective entire societies; as well as specific groups and individuals on multiple levels. Germans living in Germany may still be feeling the effects of the Holocaust today; African Americans may still be feeling the effects of slavery; the effects of the genocide in Rwanda will probably be felt by Hutus and Tutsis for generations.

d) Multi generational trauma. Birth trauma

Couples may be affected if they want but cannot have children, if they suffer from miscarriages, abortions, failed IVF attempts or the death of children at an early age,

Babies may endure trauma during conception, in the womb before they are born or after they are born if their parent does not want them, is dealing with grief from the loss of another child, is dealing with trauma from their own childhood experiences (incest, neglect, abandonment, violence), is dealing with addiction or confusion. Parents may not recognize they are subconsciously passing these traumas on to their unborn or recently born children. Ceaserean sections or drugs utilized during pregnancy, a mother's alcohol or drug use can also cause trauma. An unborn child, fetus, or baby may feel trauma associated with the loss of a twin during conception, pregnancy or birth,

These children may suffer from psychological trauma and employ mechanisms such as splitting in order to survive.

e) **Leadership roles to help people deal with trauma**

Leaders are not only those on the front lines of political and social movements, policy makers or speechmakers.

As human beings living in the modern world, one of the most powerful things we can do to create change in the world is to help each other to see. Counselors, therapists, trauma-psychologists, constellators; social workers, and other professionals work to support traumatized people to align their subconscious realities into line with objective reality. They do this by uncovering roots of trauma, thus bringing subconscious or unconscious events to consciousness.

On a collective scale, this work may also be done in other supportive capacities.

World leaders may try to bring to light collective trauma that resulted from war or segregation; to therapists or systematic constellators who bring patients individual trauma to light; to artists, musicians and writers who try and bring to light an inequity in the world.

Simply by becoming consciously aware of the roots of issues that are damaging us as individuals and a society; can move mountains. We are living during a time when people's consciousness is shifting both individually and collectively. This is due largely to the fact that people are uncovering and accepting responsibility for their own traumas, that society is making it not only acceptable but desirable to do so.

Trauma and Leadership

Bene Brown PHD, is an author and research professor who advocates for individuals to embrace their vulnerability; and admit the things that have shamed them; in order to transform the way we live, love and lead. Brown asserts that it is only through embracing our trauma and the things that have harmed us; that we can really learn to trust each other and align with each other to heal ourselves, society and the world.

Leadership may occur in less obvious ways in cases where there is no real known obvious trauma—or a lesser known trauma. People who are more inclined towards introversion by nature, may have felt traumatized or persecuted throughout their lives due to their nature. They may have spent time on the sidelines, felt like it was unsafe to talk during a conversation, railed inside because nobody listened to, respected or asked for their opinion. They may have spent a lot of their lives standing back and witnessing what is going on in the world, and honing unique skills in the process to help others.

Susan Cain, pinpoints one of these unique groups in her book "Quiet. The Power of Introverts in a World Obsessed with Talking." In her book Cain points out how many societies obsessed with action and talking and who do not tend to give much attention or status to those who are quieter, more observant or better listeners by nature---can benefit from the skills of these individuals. Introverts are great at bearing witness to individual and collective trauma, to being able to help support others in their struggles to understand these traumas, to pave the way for change by helping others learn to take a step back, and be better listeners and watchers in the world.

Again, the most helpful ways we have to shift collective consciousness is to know ourselves; to learn to understand what makes us function on a psychological level, to learn how we are triggered and react to conflict in our lives. Once we learn to do this, we can learn to check in with ourselves whenever it is necessary. There is no time that this is more necessary as when large scale events occur that may cause widespread societal trauma.

Chapter 11

Victim/Perpetrator relationships on large scales, inner and outer victim/perpetrator dynamics

In the previous chapter Franz Ruppert showed us how believing ourselves to be the victim as a method of surviving trauma, can scar us in various ways. We may align with the perpetrators by clinging to them on an emotional level, seeking to protect them, excusing their actions, and by not seeing them as they really are. We may seek to "fix" them.

Aligning with perpetrators may also cause us to become perpetrators towards others, or towards ourselves. It can cause us to become self-destructive; or turn our anger inwards which can cause physical (chronic diseases) and mental health problems, suicidal thoughts, or depression or other ailments.

Being a perpetrator can be just as wounding as being a victim. When one commits an act that hurts others they can suffer from feelings of guilt, shame or ostracism; or having a negative conscience.

Likewise, in order to cope with the damage one has done, defense mechanisms may be employed. The perpetrator may not be able to acknowledge the harm they have done to another person; he/she may feel righteous and justified for the acts that he/she committed (in lieu of feeling guilty). The perpetrator may blame the victim or feel himself/herself as being victimized. He/she may also create a scenerio in his/er brains in which the acts he/she committed are part of a larger ideology in which they see themselves as the hero. An extreme example may be a dictator who justifies his/her work committing genocide on a race of other people; by building a scenerio in which that race is inferior, dangerous and/or threatening.

Perpetrators on large and small scales

Perpetrators and perpetration come in all forms. A man/woman may be an obvious perpetrator as in the case of a murderer or rapist. Then there are other levels of perpetration. A predatory mortgage lender may convince people with bad credit to invest their life savings in homes they cannot afford.

There are less severe forms of perpetrating actions which may not even be viewed that way. Office workers who form an alliance against another worker, investing time and energy to gossiping about him/her or trying to get him/her kicked out of the office may be perpetrators.

Subjectivity. Labeling victims and persecutors

The label of perpetrator is specific to the person who is labeling him/her. Someone may believe another people to be a perpetrator that others don't. A wife may believe a husband who cheats on her to be a perpetrator. A teenager may view a parent who won't allow him/her to go to a party to be a perpetrator. In the case of the office worker, the gossipers may feel themselves justified in their actions because that worker may have done something to them they may view as persecution---for example crossed a union line or lowered their salaries by coming into the office for less money.

Becoming what we learned

Assuming other roles, Victim/perpetrator split, Victim becomes persecutor, persecutor becomes victims

The roles we assume as victims and persecutors can also be passed on. If a child is persecuted by an abusive parent; and subconsciously absorbs the ideology that parent lays down, he/she may grow up to be an abuser/persecutor to his/her child. He/she may not necessarily be aware that this is going on. He/she may be numbed towards it.

People may be neither true victims or persecutors. There personalities may waver between each role and attitude. Anger at being victimized may be turned inwards and the former victim may become a persecutor towards him/herself. This may come out in self destructive or self sabatoging behaviors, aggression towards others, confusion, distraction, addictions, depression and/or suicidal behaviors or actions

This persecution may come out in less direct ways. A man who is yelled at by his boss, may internalize his anger and come home and yell at his wife, child or dog. He/she may view aggression/depression as a normal part of living relatiionships. He she may be motivated by feelings of revenge or false atonement,

Another common reaction to victim/perpetrator splitting is the inability to establish normal intimate relationships and friendships. People may invest a lot of energy trying to lay claim to false illusions of intimacy or love. This is made more difficult if the person caught in the victim/persecutor tailspin can't come to terms with his/her past/role and/or find the ability to love and accept him/herself.

Getting Out. How to break the chain of victim persecutor in our own lives and our collective lives.

False attempts

People may acknowledge that something deep in their nature is damaged, that they have been caught in the victim persecutor cycle, be acting on triggers/habits that stem from old childhood trauma; and attempt to find a way to break this cycle.

We may first attempt to do this in earnest, but on superficial levels without getting to the root of the problem. This may include acts of revenge to attempt to annihilate the perpetrator or rebellion---struggling to fight the perpetrator without real focus or intention. We may also attempt to forgive the perpetrator or reconcile within ourselves; without really uncovering, processing or integrating our own trauma related to the victim/perpetrator mindset. There are forms of escapism that people may practice to attempt to "fix" their problems related to victim/persecutor trauma. These include physical or mental illness.

150

Although art and spirituality may ultimately be important, even crucial means of getting to the root of where your trauma comes from, and healing; people are also apt to "practice" art and spirituality on levels that do not get to the heart of their trauma. They may trick themselves or cheat themselves into believing they are healed, or have moved above what happened through techniques such as positive thinking which may gloss over, bypass or allow them to truly avoid dealing with and integrating their trauma. In these cases people may "fake it" and convince themselves they have healed, when in actuality they are just creating and exercising another elaborate defense mechanism (for real techniques of using art and inner resources for healing see chapter----).

In general, when there is real trauma, art and spirituality are most effective healing techniques when used in conjunction with more formal deep psychological practices such as constellation work.

Effective ways to overcome victim trauma. Individual

One of the toughest things for people to acknowledge is that they have been a victim. This often requires one reopen old wounds, often reliving/re-experiencing traima in order to acknowledge and accept that one has been wounded. Major complication contributing to people's denial of old trauma is the fact that persecutors often do a number on victims; causing them to "blame" themselves for what happens, to mask what happened so nobody will know, or to view what happened to them as being routine or normal. Likewise, victims often employ a number of defense mechanisms (which become habitual) throughout their lives in order to survive the trauma. These often operate on deep subconscious levels, and as time passes it becomes more and more difficult to see and get rid of them.

Victims must learn to admit how they feel on deep emotional levels about what happened to them. They may cycle through emotions such as rage, shame, powerlessness, sadness confusion; for a while. After a while, these emotions harm us less. We learn to feel them, and then we can let them go. After acknowledging these emotions, even to the point of courageously re-experiencing them; they may make us feel sad in the future but they can never harm or break us again.

After admitting victimization, people must learn to feel compassion for themselves and what they have endured. This is different from self-pity, which may be necessary to experience for a time, but can ultimately become crippling and cause us to remain in the victim role. It is more like self-empathy. Imagine how you would feel if you witnessed somebody going through what you went through, how you would want to make them feel that you cared about and could understand what they went through; and that you want them to heal.

Ultimately, learning to overcome the victim attitude requires one to move past their victimization to recover their personal power; and to create healthy alternative ways to interact in the world. This includes recognizing one's own basic goodness, establishing good boundaries and establishing methods for taking care of ones damaged self.

This also requires we acknowledge, accept and seek to alter ways in which we may be emulating persecutors behavior and becoming persecutors in our own lives.

Effective ways to overcome the perpetrator attitude-- Individual

It is equally, if not more, difficult to acknowledge ways we have acted as persecutors. It is helpful if we view our actions simply, in context, and acknowledge that there is no clear black and white, right or wrong view in life in general. We all act out different roles, and we all have the power to rid ourselves of undesirable roles.

The first way to overcome the perpetrator action is to acknowledge things we did that may have harmed or victimized others. If these are difficult acts to acknowledge for example abuse; we might also acknowledge that it is an act of bravery to do so; and that just by acknowledging it we are shifting things, as we also pave the way for the healing of the victim we harmed.

We must also "feel it" the guilt and responsibility for what we did, our anxiety about acceptance. We must seek to find compassion, and then empathy for the victims.

We might take concrete actions to make amends with victims; in some cases offering an apology or compensation (monetary or in some other form). Franz….notes that persecutors should resist urges to make a lifelong atonement for what occurred.

Chapter 12:
Collective Victim and Persecutor mentalities and how to overcome them through leadership

Victim/persecutors on a large collective scale.

We have seen throughout this book, how we can unwittingly pass on trauma, baggage from our ancestors, parents, selves and children; how trauma can be a chain, that if unacknowledged can continue on and on.

Similarly, this chain can be recognized on a collective scale. When incest and sexual abuse become the norm in a society; violence is also bound to be turned outwards. These may include obvious forms of violence such as murders, drugs and human trafficking, or political persecution of a group. It may contain less obvious forms of violence such as economic violence. Current economic systems which are driven purely by human greed/selfishness, where people do not exercise integrity in their economic dealings, where exploitation becomes an acceptable norm (getting ahead at the expense of the other, get them before they get you mentality); are obviously forged by a group of individuals who have been victims of persecution and/or are prone to violence. The effects of living in a society in which violence and persecution are inherent on subconscious and unconscious levels are also obvious in factors such as the extreme depletion of our natural resources and the poisoning of our natural environment, extreme gaps between rich and poor/have's and have nots, civil unrest in the form of protests, school shootings, mass executions and general terror. One would expect if we were living in more equitable societies where people trusted each other more, if the victim/persecutor split wasn't motivating us, we would see healthier, more balanced results. Franz....notes: That the victim/persecutor split results in the traumatization of bonding systems, and entire societies which are dominated by trauma.

We can also see the residual effect of large scale persecution in extreme examples of political violence. Populations who have been surrounded by violence in wars or genocides may go on to become exploiters of other races or populations; particularly if they were persecutors when they were young before their brains had a chance to fully develop. Children who grew up with parents in prison, or in tough neighborhoods where violence is the norm; may go on to wind up in prison themselves.

a) Alternative ways to heal Collective Traumas.

When our psyches are split, as individuals and collectives, terrible things can occur. Human beings may be demoralized, terrorized, oppressed due to their race, culture, or socioeconomic status.

Traditionally, systems settle old scores. These might be the politicians attempting to reconfigure an old systems (for example attempting to install a new leader after a dictator has been overthrown, or to try and institute democracy in a formerly communist society). Lawyers mediate disputes whether they be criminal, business or family (as in the case of divorce).

Of course, this brand of meditation does not always work. When the US went into Iraq purporting to attempt to overthrow Saddam Hussein and set up a new democratic system for the Iraqi people, they failed miserably. This happened because they failed to take into account the thousands of years of Iraqi history, their cultural realities, or even their wishes. Men, women and children are sentenced to prison terms that essentially keep them out of society; yet they often recycle back into communities where they commit similar crimes again. Prisons abound with men, women and children who committed crimes stemming directly or indirectly from childhood abuse or trauma. In these cases, mediation is not effective.

Alternative conflict resolution methods abound. Alternative Dispute Resolution is an alternative means of people or systems in dispute to come to a resolution without going to court. Some methods include negotiation (where there is no third party), mediation (where a third party the meditator facilitates the process but does not have power to decide what happens to either party) and collaborative law—through which lawyers intervene with preestablished standards outside of court.

There are programs which are developed which go beyond alternative conflict resolution methods. Restorative and community justice programs are currently being created through which men, women and youth who have committed crimes are not sent to prison but are given the opportunity to make amends to the families and communities they hurt through service projects. India has a system called "Lok Adalat" or people's court through which elders help resolve conflict within communities. Several tribal communities in Alaska have similar village court systems through which elders hear cases and dole out community service sentences in response to offences or crimes.

The latter is more effective because it goes to the root of the problem, the heart of the community and acknowledges on a deep level the humanity, standing and motivations for both victim and offender. It helps to heal trauma for both the community and the offender---by delving into its core. System wide processes tend to undermine the humanity of the situation. When people are warehoused away from the places where they committed crimes; the offender does not have ample opportunity to heal or make amends; and the victim and/or victim's family does not have the opportunity to forgive.

Any leadership we install to help create real change in the world, needs to delve deeper; to get to the root of individual and collective trauma, to dig out the decay and detritus and to clear the way for new growth to occur. The most effective way we can do this is to apply constellation work and transgenerational psychology on both individual and collective levels. We must seek to bring to light what ails us, in order to transform ourselves and the world. This can pave the way for more effective, systematic leadership to occur.

These kind of techniques are necessary in the case of large scale trauma or systematic violence, for example the case of genocide. When World War II occurred, many Christians who lived in villages where the

Nazis were in power, may have watched or contributed to the murder of their neighbors. The Jews whose families were exterminated may have found it difficult not to blame the perpetrators, may have found it difficult to forgive. Yet generations later---Germans and Jews who lived through the war together are finding ways to reconcile. Forgiving perpetrators can have massive healing effects on entire communities, societies and on the world. It is only when we acknowledge these traumas, acknowledge the effects they had on our ancestors and how our ancestors unwittingly passed those traumas down to our grandparents, parents and ultimately to us. These traumas may not be easily traced back to the source. It may take us some time to understand that the physical abuse our father committed on us, the abuse that we then went on to commit on our children through verbal abuse before we knew any better; is rooted in the trauma/pain of our grandfather who was forced to kill people of a different race during a war. It is only through this acknowledgement, through this unrooting, that we can find the means to break the cycles of violence and truly begin to heal.

We may also stop blanket large scale atrocities from subconsciously affecting our interactions with others on subconscious levels. For example On September 11, 2001 two planes were flown into the World Trade Center, and thousands of people were killed. The media inundated us with images of the destruction, and later with images of Muslims in Brooklyn cheering which caused many people to associate all Muslims or anyone they believed to be Muslim with the bombing. Others strove to look deeper. They did not associate the terrorists who flew the plane with their neighbors or men and woman who walked the street in their cities.

When terrorism, war, genocide, hunger or natural disasters occur, or we are threatened with their future occurrence--it would be in our best interest to learn techniques to help ourselves and others disseminate information, to identify and/or alleviate blanket political or cultural wounding or scarring as it might be occurring. We must learn to understand how we function in society on the deepest levels.

Getting Out. How to break the chain of victim persecutor in our collective lives.

Victims healing on collective levels

A remarkable thing happens when people begin to acknowledge the "terrible, shameful" ways they have been victimized. They come to recognize that there are resources out there to help them, guidance about how to heal; and perhaps just as if not more importantly that they are not alone. When one person begins to heal, they often find others that are healing from the same types of wounds. Groups of survivors may be found--for example in the various forms of Alcoholics anonymous or narcotics anonymous groups (including groups for family members of alcoholics or drug abusers); incest survival groups

When people begin to heal from these things, individually, and in collective groups; they become more and more acknowledged publically. Years ago, there was very little public acknowledgement of the incest or molestation of children that occurs at a rampant rate in society. There was no mention of it on television, or in the newspapers. Accordingly, the court systems did not support children who were molested. Sentences for abusers weren't severe, abuse was difficult to prove, and victims were often "revictimized" by court/legal and hospital systems when they came forward. Children were often viewed as liars or guilty; and the abusers often were free to travel through society largely invisible, often recommitting their crimes on more children. Parents of children who were abused by their spouses or other adults; often did not know where to turn or how to regard what happened to their children. They sometimes viewed it as the child's fault.

Times have changed. It is now common knowledge that children are molested, that adults commit incest. More victims are urged to come forward. Television programs across the board (talk shows, news programs, drama's, reality television) bring light to the fact that these things happen, that they are wrong, and that people who know victims should come forward and report the persecutors. Likewise, there is more therapy and groups for children that have been abused/molested. The courts have more support mechanisms in place for abused children; abusers receive stiffer sentences. There are registry's where abusers must go on public record about what they have done; they can no longer go through society invisible. Therapy and other measures are mandated to try and ensure they will not harm more children. It is no longer an invisible problem. We have made huge collective leaps in acknowledging that abuse happens, that it is wrong, and that we should make efforts to eradicate it from our realities.

Sometimes, leadership is realized because we are willing to witness the trauma inflicted on ourselves or other people; to give voice to that trauma, to join ranks with others who have been victimized or support those that have been victimized; to step up and bring to light a problem that needs to be quashed from our reality---a massive change that needs to occur. When enough people join ranks in this way: when we uncover the unacknowledged roots of individual and collective victimization; when we set an intention to acknowledge and heal from this victimization as individual and as a collective; we are shifting consciousness, bringing the subconscious to conscious light, paving the way for real lasting change to be created in the world.

Persecutors healing on societal levels

In this world, as human beings it is sometimes easy to blame persecutors. It is tough to see persecutors as former victims themselves, to acknowledge that some of their behavior is generally learned.

We are living in a transitional time in the world in the sense that people who were born in the past few generations; are finally seeking to break chains of violence---a lot of which is rooted in ancestral and collective trauma from a time that has passed. This is the age of admission. Psychological treatment in many forms (including constellation work) is becoming normalized. People are urged to seek professional help, to deal with their "problems" and issues. Seeking help is no longer viewed as a shameful thing.

Likewise, we are being urged to admit our collective trauma; to urge persecutors to admit their actions publically and to make amends for them. In this way our collective consciousness starts to shift.

There are programs through which persecutors are humanized in ways they never were. Cycles of violence which lead to the mass incarceration of young people of color from inner cities are being acknowledged. People are acknowledging that mass incarceration leads to cycles of recidivism, and does not work to deter crime. Programs are developed in which families of people who have been murdered are finding ways to forgive the murderer. They are spearheading alternative restorative justice programs in lieu of punitive measures.

Likewise, atrocities that were committed centuries ago are being publically acknowledged. People who committed war crimes, helping the Nazis, Albanians, Nazis, Contras to brutalize, torture and kill people are being brought to justice. Watchdogs like Amnesty International have been established to safeguard human rights that were blindly violated in the past. A good example of this is the situation that took place to safeguard the rights of protesters in Ferguson Missouri during protests following the shooting of an African American boy by a police officer.

Global response and increased awareness of the problems of human trafficking and sex trafficking are also good examples of people holding persecutors accountable on a societal level and shifts in collective consciousness through attention to the issue. Media attention, legal attention and protests/movements have been focused on holding persecutors accountable and attempting to break down the intricate system and chain that holds human beings as slaves across the world.

Another good example of collective response to perceived persecutors is the Occupy Wall Street movements in the United States which began when people joined forces on September 17, 2011 in the Wall Street area of Zoccoti Park; to protest economic and social inequities in the world. The movement drew it's inspiration by Spain's anti-austerity protests from the 15-M movement. The movement continues today with various groups springing up in different parts of the country and the world to protest the extreme inequities in wealth distribution (1 percent of the country holding the majority of the wealth) and the extreme influences of corporations who are not always held accountable for their actions and who hold what is perceived to be too much influence on the government and media.

A Word About Collective Response to Trauma

Just as human beings can make "false" or superficial attempts to overcome trauma and victim/perpetrator mentality; so can groups through movements and protests if they are not paying attention. An institution or individual may be scapegoated as a "persecutor" and a system set up to be overthrown, without proper attention being given to the root causes of the problem.

On the flip side, movements, protests, and massive shifts in collective consciousness which result from increased knowledge of a system or group that has been functioning as a persecutor and victimizing individuals or groups; which functions in an integral fashion by taking account of the root causes of the problems, its effect on individuals and groups is miraculous and has the potential to create real and lasting change.

Leadership potential in revamping individual and collective systems perpetuating the Victim/Perpetrator mentality

The mechanisms, structures, slogans and policies we adapt during this time can really affect change.

For example, when terrorism occurs, people join forces and adapt new mechanisms to resist it. There are leaders who pave the way for this to occur. Again, we can look towards Malala Yousafzai, the young Pakistani girl who wrote a blog when she was 12 years old detailing her life under Taliban occupation, and advocating for girls to go to school. Although Malala was shot by the Taliban on a bus, and nearly lost her life, she went on to become an outspoken activist for girls and womens rights to be educated and became the youngest person to win the Nobel Peace Prize.

Slogans can also pave the way for change. The phrase "Je suis Charlie" (French for I am Charlie) was a slogan adapted after twelve employees of the satirical weekly newspaper Charlie Hebdo, were massacred. The phrase became widely associated with freedom of speech and the press and as a device to resist terrorism, oppression and threats to freedom of self-expression.

Chapter 13:
Preparing our children to be future leaders.

Parenting and shame

In Brene Brown's Book, I thought it was Just Me, she notes how her interviews with women revealed they often felt judged, criticized or shamed by other women at the way they raised their children. She notes that women are often shamed when it comes to the way they raise their children; where as men are often held to similar forms of scrutiny when it comes to their financial or professional success or non success.

It is tough to be a parent. There are no instruction manuals. Parenting is more challenging in the modern era where parents often must work long hours in addition to their parenting; often simply to put food on the table and make sure their children are fed and they pay the rent. The extended support systems that were once in place through extended family and neighbors; are often replaced with child care facilities and schools. Advice and guidelines for parenting on topics ranging from health to discipline are as likely to come from popular culture---ie, television news and radio programs; as they are from relatives or friends.

With all this uncertainty, it is likely that judgment might occur.

Brown notes that judgment begats judgment---that fear of being judged also causes us to judge others; and that it is nearly impossible to fake being "non judgmental as it appears in our expressions and body language. In order to break this chain, she suggests we be more mindful of what we are thinking, feeling, seeing in regards to parenting; and where our opinions are really rooted. Of course, a good technique for doing this is to go back to our own childhoods, again, to uncover the proverbial skeletons in our closet, to bring to conscious awareness our subconscious motivations, to root them out. This will help us both in interacting and forging communities and trust with other parents; and in our own parenting. Judgment does not necessarily come from the outside; but to the critical voices of our own parents and peers from the past I our heads.

Parenting and Childrens image

Body Image

Brown also notes that parents are apt to pass on shame to their children as it relates to self image, particularly body image and attractiveness for girls, and that this shaming can cripple a child's spirit and self esteem; and is one factor that can prevent them from being strong leaders and role models for others.

Boosting a child's self esteem by helping them to affect positive and realistic body and appearance images; by emphasizing for instance health over weight loss, or inner beauty in conjunction with outer beauty can help them to be less bogged down with narcissitic and neurotic expectations of themselves; and more fully capable, self reliant human beings in the future.

Brown notes that body shaming has far reaching impacts into how we love, live and parent.

Parenting and perfection

When we are parenting, we should always cut ourselves a little slack. Children are difficult, and we are fallible human beings. There is no such thing as perfect parenting. There is no sure fire recipe that will yield a perfect little human being. We all learn from our mistakes---children and parents. Imposing unrealistic expectations on children whether it be for their academic achievements, friendships or behaviors can cause great harm; often having the opposite effect as we desired.

Brown notes that we are particularly apt to attempt to "shame, terrify or judge" our children when they are not living up to our expectations of how we would like to be perceived as parents, when they do not live up to our ideal "parenting image." It would behoove us to attempt not to inflict that on our children, to acknowledge they are independent human beings that do not "belong" to us and do our best to allow them to live out their personalities and roles in the world. We must of course exercise proper judgment when it comes to parenting, set guidelines, restrictions and expectations for behavior and achievements; but it is best to keep these realistic and always acknowledge our child's uniqueness and independence from us.

When things go wrong. Parenting and shame

There is no perfect human being. We learn as humans through our mistakes, through a series of unique trials and tribulations through which we experiment and often "mess up." The ways in which we mess up are not necessarily exclusively disasters; because they teach us and provide us perspective through which we measure other trials and tribulations in our future lives.

Parenting can be toughest when our children are growing and we recognize they are in serious trouble. We may have a child who grows up depressed or with mental health issues. We may have a child who has issues with addiction. We may have a child who winds up in an abusive relationship or being an abuser. We may have a child who has problems with money. We may have a child who robs a bank.

Some of the toughest things to do in this situation are to take a step back and allow our child to mess up. It is tough to resist trying to fix it, blaming ourselves; blaming, shaming or passing judgment on him/her.

Sometimes, tough situations require we exercise tough love--- allowing our children to know we are there for them and support them and will be there if they are willing and ready to make a shift in their life; but that we do not necessarily agree with or support their behaviors. Sometimes loving someone requires we take a step back and allow them to come back to themselves.

Sometimes, our children's troubles are the result of our own troubles. This can be hard to see, and to admit. Our aggression or addictions or bad choice in partners may have contributed to our child's depression, violence or addictions. If we know this to be the case, sometimes the bravest thing we can do is to sit down with him/her and admit what we know and what we have done; to apologize if necessary, to offer to help him/her with therapy or other support networks to get help.

Again, with parenting comes great responsibility. Parents in this generation are lucky in a sense because we have come to a place where we are beginning to deal with our own trauma, and to attempt not to pass these traumas down to our children. Our children are growing up with more support to deal with their own trauma; and are less likely to develop habitual reactions to trauma and to have outlets that keep them from being re-traumatized.

Children who have dealt with their trauma or who have less trauma due to the fact that we are aware of our own---are more likely to grow up to be powerful, independent human beings who trust and love others. They are more likely to be affective guides and leaders for others.

Shifting the shame blame paradigm

Brown notes that the influence parents have over their children's ego, self- esteem and sense of potential is boundless. It never goes away. Likewise, if parents have unwittingly and unintentionally used shame, blame, ridicule or judgment as a means of trying to control, motivate or influence behavior; these feelings might stick. More often than not, these are habits we learned from our own parents, that we carried with us on a subconscious level our whole lives. Again, awareness and ownership of our own problems and traumas, can pave the way for our own healing and for our child's healing.

The good news is we are with stuck with our children for the length of their lives and they are stuck with us. We have plenty of time to make things right, to heal our relationships. Brown notes that simply acknowledging some real or perceived wrong we may have done to our child---can have an extraordinary impact. For example simply saying "I understand how my actions made you feel," or "I'm sorry," without overanalyzing or defending can really help.

Resilience in children

The bad news is we can really mess our children up. The good news is we can only mess them up to a point. Children are remarkably resilient. They are also hard wired to be proud of who they are, and to know who they are.

We can also teach our children to increase their resilience, particularly in the face of adversity or hatred, blame or disconnection.

The best thing we can do with parents is to attempt to raise our children with the qualities we cultivate in our own lives. These may include things like empathy and compassion for other being, being of service to others, bravery in the face of adversity or despite fear, and a sense of capability and pride in who they are and what they can accomplish. We can consciously choose to build these qualities in our children. We can teach them to connect with others, to see beauty, to love and to trust. We can teach them that they belong in the world, and that the world is not such a scary place.

Children and connection

One of the biggest problems with today's society is the dissociation and isolation endemic to it. People are taught not to trust each other, to be terrified of the world and the people in it, that there is not enough to go around. Children learn these values from their parents. If we alter our realities so that we are more connected emotionally, intellectually and physically to others---we will also be altering our children's realities. If we teach them to rely on others as well as themselves, to teach others as well as themselves, to help others as well as themselves---we will be paving the way for a shift of consciousness and change in the world.

One of the most important things we can do is to teach our children that they are rooted in the earth, that they belong here, that we all belong. There are unique gifts and roles we all have to be celebrated. The earth is a rich, abundant place with plenty of resources to go around. Our children must be taught how integral those resources are to them, and how fragile; and to respect and defend them. We all have a right to be here. We all have to stand together to be here, to shift the consciousness of the world and to heal ourselves and each other. It is only when we stand together, when we all admit we belong, that this kind of change can occur.

Embrace your child's unique gifts

Artist, introert, leader,

When it comes to your child's talents, gifts or roles in the world; it is bravest to watch them and let them live out their roles in the world; without trying to impose your own expectations or value system. Support and nurture the talents they have; acknowledge things that are difficult for them not necessarily as weaknesses or obstacles that they have to overcome and do what you can to help them to plug along. Help your child come into his/her own, but allow him/her to lead the way.

This society can often be perceived as self destructive. There are clear cut disciplines and roles that we tend to give more respect to or credence to, such as earning money, being outgoing, talking a lot or loudly or in a convincing fashion, being overly social, being good in business. There are roles we sometimes tend to give less credence to; such as creativity, musicality, introversion, thoughtfulness, what is perceived of as passivity. Things are changing and we may come to realize we need more of these qualities in the future. There are roles children are being groomed for we may not even know exist yet; such as an extraordinary aptitude at technology, making use of other sensory perceptions for example future chefs with a more pronounced ability to taste the nuances of food or future therapists with a more advanced ability to read or empathize with others.

In Susan Cain's book, Quiet, The Power of Introverts in a World That Can't Stop Talking, she implores parents of introverted kids to embrace their childs introverted qualities; and that while parents of introverted children may worry that they will not be accepted in or fit into society; it would behoove them to note that they are probably not going to be able to change their fundamental nature. Likewise, she notes, those introverted qualities may be developed during the course of a child's life and yield good results such as a greater capacity for self-awareness, thoughtfulness, observation skills, the ability to listen to and empathize with others on deeper levels. Although it may be difficult or troublesome for parents to accept these qualities in their children; it would be in their best interest to take a step back and see what the world looks like from the perspective of an introverted or quieter child.

She also suggests that some of the commonalities shier children might exhibit is the need for more down time or quieter environments, breaks between activities, and perhaps slower but deeper levels of conversation. They may need

Whatever your child's gifts are, embrace them. If they are boisterous and active and engaging, but have trouble focusing on a single task try and find a multitude of activities or sports they enjoy to keep them busy. If they like making a mess, introduce them to an art project. If they are orderly and overly organized to the point of seeming a little OCD---give them a recipe book to catalogue.

It is also good to ease them into situations that are uncomfortable to them. If a child is extremely introverted to the point of extreme isolation, it may be good to introduce them to other children and groups. If your child is an extrovert to the point of distraction, it may be good to attempt to introduce them

a) Leadership through systemic teaching and education.

b) **Embracing our children's connection to the unseen world, and preparing them to be future leaders.**

c) Cultivating cross cultural diversity. From shame and rootlessness to belonging.

We all belong

Children and leadership

There are several ways we can prepare children to be the kind of leaders we will ultimately need in the world.

One of the most important things we can do is to change ourselves, to work to break the cycles of trauma that have existed in our lives; and not to pass it on to them. We can try our best not to shame, judge, or criticize them; and if we do to acknowledge this and to acknowledge the ways we make them feel. We can try and embrace their talents and gifts and unique nature; to teach them to be proud of who they are and to love.

We can prepare the ground for the kind of world in which they will grow. This includes introducing them to many different groups of people and teaching them to celebrate cultural, racial and economic diversity. We can teach them how important it is to care for the natural resources of the world. We can serve by example: pointing out where injustices, inequities or violence exists in the world---and the importance of joining movements protesting or resisting these things. We can teach them to trust themselves; to connect with, trust and love each other. We can whenever possible, follow their example, learn about the world they are going to inherit through their actions and gifts unique to them and their generation.

"Everyone has inside of him a piece of good news. The good news is that you don't know how great you can be! How much you can love! And what your potential is." Anne Frank

www.ingramcontent.com/pod-product-compliance
Lightning Source LLC
Chambersburg PA
CBHW061305220326
41599CB00026B/4733